# Leader Development in Army Units

## Views from the Field



Peter Schirmer, James C. Crowley, Nancy E. Blacker, Richard R. Brennan, Jr., Henry A. Leonard, J. Michael Polich, Jerry M. Sollinger, Danielle M. Varda

Prepared for the United States Army



RAND ARROYO CENTER

The research described in this report was sponsored by the United States Army under Contract No. W74V8H-06-C-0001.

**Library of Congress Cataloging-in-Publication Data**

Leader development in Army units : views from the field / Peter Schirmer ...
p. cm.
Includes bibliographical references.
ISBN 978-0-8330-4200-2 (pbk. : alk. paper)
1. United States. Army—Officers—Training of. 2. Command of troops.
3. Leadership—United States. I. Schirmer, Peter, 1970–

UB413.L43 2008
355.3'3041—dc22

2008006838



Published 2008 by the RAND Corporation
1776 Main Street, P.O. Box 2138, Santa Monica, CA 90407-2138
1200 South Hayes Street, Arlington, VA 22202-5050
4570 Fifth Avenue, Suite 600, Pittsburgh, PA 15213-2665
RAND URL: http://www.rand.org/
To order RAND documents or to obtain additional information, contact
Distribution Services: Telephone: (310) 451-7002;
Fax: (310) 451-6915; Email: order@rand.org

# Preface

As an institution that promotes lifelong learning, the Army strives to ensure that its leaders are immersed in a learning environment throughout their careers. Officers, for example, move back and forth between the Army's school system, which teaches job-related skills and knowledge, and operational assignments, which teach personal aspects of leadership, provide experiences and immediate feedback that drive self-development, and expose leaders to role models and mentors. While it is widely believed that experience in a variety of assignments makes a large contribution—possibly the most significant contribution—to the development of Army leaders, little in the way of organized Army-wide unit-level leader development programs exists. Furthermore, the operational commitments in Afghanistan and Iraq have increased the demands on units, making leader development programs more difficult to design and carry out. The Center for Army Leadership asked RAND Arroyo Center to help the Army in identifying effective and feasible unit leader development programs. This document reports on the results of those efforts. It should interest those involved in Army leader development and personnel management.

This research has been conducted in RAND Arroyo Center's Manpower and Training Program. RAND Arroyo Center, part of the RAND Corporation, is a federally funded research and development center sponsored by the United States Army. Questions and comments regarding this research are welcome and should be directed to the leader of the research team, Pete Schirmer, at Pete_Schirmer@rand.org.

The Project Unique Identification Code (PUIC) for the project that produced this document is ATFCR06020.

For more information on RAND Arroyo Center, contact the Director of Operations (telephone 310-393-0411, extension 6419; FAX 310-451-6952; email Marcy_Agmon@rand.org), or visit Arroyo's Web site at http://www.rand.org/ard/.

# Contents

**Appendixes**

# Figures

# Tables

# Summary

Army leaders believe that a very significant contribution to their leader development comes from their experience in operational assignments. Yet there are few studies that indicate whether Army units even have leader development programs, and if they do, what the programs consist of and how well they are executed. The Center for Army Leadership (CAL) asked RAND Arroyo Center to help the Army identify effective and feasible leader development programs in operational units.

## Hundreds of Officers Participated in the Study

Over 450 officers met with teams of Arroyo researchers during the summer and fall of 2006 to discuss leader development in Army units. The Arroyo teams met with officers at the Army War College (AWC), National Defense University (NDU), the National Training Center (NTC), the Command and General Staff College (CGSC), and five captains' career courses (CCCs). The CCCs that were chosen represent combat arms (armor, infantry, and engineer), combat support (chemical), and combat service support (combined logistics) branches.

A total of 405 officers up through the grade of major completed a written questionnaire that inquired about the leader development activities in their last operational assignment. After completing the questionnaire, those officers participated in half-hour discussions with the Arroyo teams. The 61 participating colonels and lieutenant colonels did not complete a written questionnaire but participated in longer discussions, often lasting two hours. During these discussions we shared some of the results from the questionnaires completed by the more junior officers. Table S.1 provides a summary of the source and rank of the participating officers.[1]

**Table S.1**
**Number of Participating Officers by Experience Level and Source**

| Experience Level | Source | Number of Participating Officers |
|---|---|---|
| Junior Captains | Armor CCC | 108 students |
| | Infantry CCC | 92 students |
| | Chemical CCC | 14 students |
| | Engineer CCC | 25 students |
| | Combined Logistics CCC | 43 students |
| | *Subtotal* | *282 junior captains* |
| Majors/ Senior Captains | CGSC | 74 students |
| | NTC | 29 O/Cs |
| | Armor CCC | 10 small group instructors |
| | Infantry CCC | 9 small group instructors |
| | Chemical CCC | 1 small group instructor |
| | *Subtotal* | *123 majors/senior captains* |
| Colonels/ Lieutenant Colonels | NDU | 13 students<br>1 faculty member |
| | AWC | 19 students |
| | NTC | 8 O/Cs |
| | Assigned to DC area | 20 officers |
| | *Subtotal* | *61 colonels/lieutenant colonels* |
| | **Grand Total** | **466 officers** |

[1] The written survey did not ask officers to provide their rank, because we assumed that all students at the CCCs would be captains. However, we did observe among the respondents a few lieutenants (fewer than ten). The responses of these officers are included in data characterized as coming from the junior captains, because their presence at the CCCs indicated imminent promotion to captain and they had experience very similar to that of their fellow students.

Participating officers had extensive and recent operational experience. Almost all of the junior captains—the students at the CCCs—were platoon leaders in their last assignment, and a large percentage were deployed. Most of the majors and senior captains were company commanders in their last assignment, and, again, a large percentage were deployed. Almost all of the colonels and lieutenant colonels were recent brigade or battalion commanders, and half had deployed.

Twenty-three of the junior captains and eleven of the majors and senior captains had last served in a National Guard or U.S. Army Reserve unit. Their questionnaire responses and discussion comments were very similar to those of the active component officers. All are presented together in this monograph. Separating the responses of reserve component officers from those of active officers would have minimal impact on the tables and figures and no impact at all on the conclusions and recommendations. Nothing in the data gives reason to believe that the leader development activities within reserve component units are significantly different from those in active component units, although a larger sample of the former would be required before one could draw definitive conclusions.

## There Is No Standard Leader Development Program

All units conduct training on a host of collective and individual tasks, but activities intended to develop the broader range of leadership skills vary greatly in content, frequency, and perceived quality. In short, there is no set of activities that could be characterized as a standard or typical unit-level leader development program. Table S.2 shows the percentage of officers who said they participated in various leader development activities in their last operational assignment. Even a required activity, such as keeping a written self-development plan and reviewing it with the rater, is not done consistently across units. In addition, leader development programs are not all-or-nothing: Units that do one type of activity, such as prescribing required reading, do not necessarily do another, such as conducting staff rides. The variation is due to several factors—unit roles and missions, location, geographical dispersion, unit readiness, and, most important, the unit commander.

**Table S.2**
**Percentage of Officers Who Participated in Various Unit-Level Leader Development Activities**

| Activity | Majors/ Senior Captains | Junior Captains |
|---|---|---|
| Commander discussed requirements of upcoming assignments and expectations to a very great extent | 23 | 16 |
| Rater discussed leadership skills at least once a quarter | 60 | 69 |
| Senior rater discussed leadership skills at least once a quarter | 34 | 40 |
| Commander encouraged officers to find a mentor or to mentor others | 52 | 51 |
| Required to have a written self-development plan | 28 | 39 |
| Required to review self-development plan with rater(s) | 21 | 30 |
| Leadership lessons generally embedded in training | 68 | 75 |
| Led or participated in an OPD class focused on leadership at least quarterly | 49 | 56 |
| Required to do professional reading focused on leadership qualities and lessons | 28 | 41 |
| Unit had a reading list | 30 | 33 |
| Unit conducted a staff ride off base | 46 | 34 |

NOTES: Numbers show percentage of *officers*, not percentage of *units*, who participated in these activities. OPD = officer professional development.

## Besides Actual Experience, Role Models and Personal Interaction Are Most Valued by Junior Officers

On the questionnaire, the majors and captains were given a list of twelve activities and were asked to select the three that were most effective in developing them as leaders. Both groups ranked the twelve items almost identically. Figure S.1 shows how junior captains ranked the items. Items at the top of the list can be lumped into two categories—operational experience and personal interaction. These are clearly more valuable to the officers than are the activities that might comprise a formal leader development program, such as staff rides or reading lists.

## Unit Commanders Are the Key to Effective Leader Development

Junior officers commonly have more than one battalion commander, and certainly more than one company commander, during a three-year assignment. The changes in command can have a profound effect on the content, frequency, and perceived quality of leader development activities. Unit leaders—especially battalion and squadron commanders—have a significant influence on the development of junior officers.

Unit commanders' decisions about what to do for leader development are constrained by a number of factors, including deployments, the roles and missions of the units, unit location and geographical dispersion, readiness levels, and the amount of individual and team experience. To respond to these factors, a number of senior officers said it would be helpful to have a flexible "tool kit" of leader development ideas.

More important, unit commanders affect leader development as role models, mentors, and counselors. As Figure S.1 shows, these are among the factors junior officers believe are most effective for leader development. Still, there is a widespread perception among junior officers that the quality of this personal interaction varies greatly depending on the unit commanders' personalities and their capacities to develop leaders.

**Figure S.1**
**Junior Captains' Ranking of Effectiveness of Leader Development Activities**

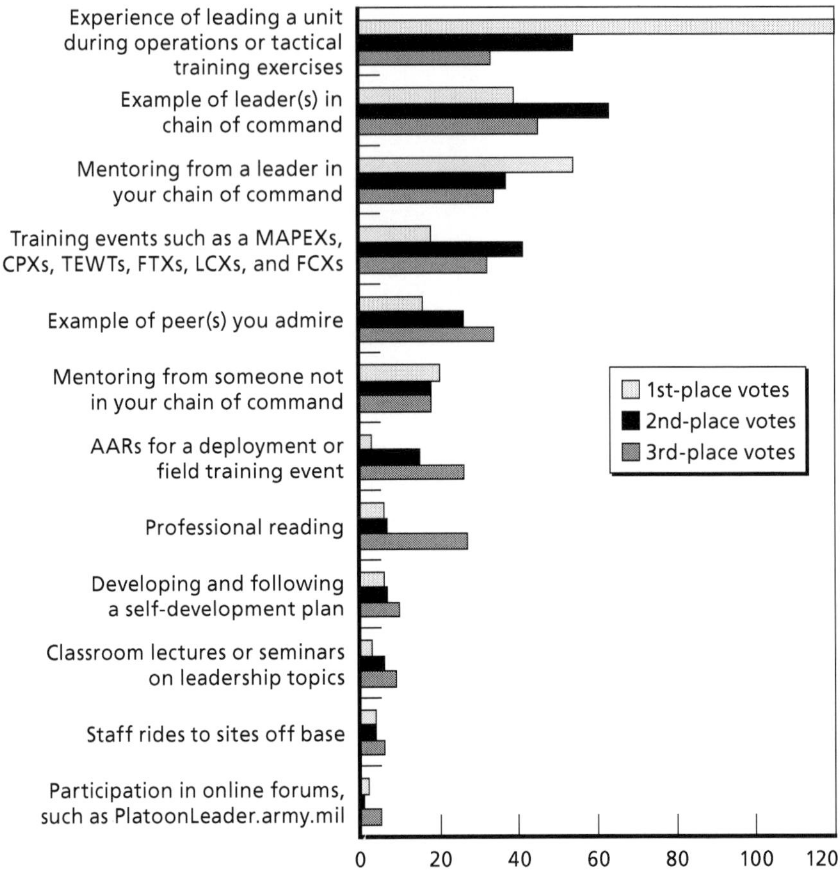


NOTES: MAPEXs = map exercises, CPXs = command post exercises, TEWTs = tactical exercises without troops, FTXs = field training exercises, LCXs = logistical coordination exercises, FCXs = fire coordination exercises, and AARs = after action reviews.
RAND *MG648-S.1*

Because of the enormous effect of unit commanders on leader development activities, the questionnaire instructed respondents to "answer the rest of the questions about your experience *within the command climate that most strongly encouraged* leadership development within your former unit."[2] To the extent that respondents were able to make this distinction when completing the questionnaire, their responses present a best-case scenario of how often activities take place and how much they focus on and affect leader development.

## Counseling Is Inconsistent

The colonels and lieutenant colonels disagreed somewhat with the majors and captains about the value and frequency of counseling, coaching, and mentoring in operational assignments. Many senior officers insisted that counseling, coaching, and mentoring occur more often than the junior officers think, perhaps because they frequently take place on an informal basis. The senior officers may have been conscious of the example they set while they were brigade and battalion commanders and would consider that a form of mentoring or counseling, whereas junior officers might not. But junior officers do recognize that informal interactions are a forum for counseling or mentoring and provide valuable developmental opportunities. Furthermore, even the senior officers frequently said they had not received high-quality counseling more than a few times throughout their own careers.

What is not in dispute is the fact that the Army's Developmental Support Form (Department of the Army [DA] Form 67-9-1a) is not widely used as required. Many of the junior officers said they had never seen one, or had filled it out on their own but never reviewed it with their rater or senior rater. But a number of unit commanders create their own developmental support forms for use by the junior officers in their units.

Senior as well as junior officers generally had difficulty describing what a junior officer's self-development program should entail. But junior officers desire to improve their leadership skills, and many mentioned informal efforts such as role modeling, self-evaluation, and reflection. More consistent counseling, part of an ongoing process

---

2 In this document, all emphasis in quotations from the survey is from the original.

of leader development, is another important way to encourage self-assessment and growth.

## Recommendations

Imposing formal programs, new forms, or reporting requirements on unit commanders is unlikely to be beneficial. Instead, our recommended strategy is to build on a process that already takes place, in which officers learn to do leader development from role models and peers. We were told by many officers, across all ranks, that people "need to see what right looks like." This approach is the basis for our two key recommendations.

### Use TRADOC Institutions to Raise Expectations for Leader Development in Units

Unit commanders, one senior officer said, need to leave the schoolhouse with ideas of how to do leader development. Counseling is one area in particular where Army schools can address several deficiencies in the current system. Counseling should emphasize adherence to a formal process with a fixed schedule for counseling sessions; instructors should help students prepare a developmental support form in anticipation of upcoming duties; and instruction should emphasize the developmental—as opposed to the administrative—aspects of counseling subordinates. Everything should reinforce to the students that this is what they should expect of themselves and their raters when they are in operational units.

The keys to this approach are teaching students through example and establishing expectations for behaviors that embrace the idea of the Army being a teaching as well as a learning organization. If instructors and faculty ensure that students are exposed to formal, personalized, developmental counseling, officers will be more likely to take their experience and expectations back to the unit and teach others by example. This is what happened when U.S. Army Training and Doctrine Command (TRADOC) institutions—the maneuver combat training centers—introduced after-action reviews (AARs). The same could happen with leader development by showing officers what right looks like in the schoolhouse, and by exposing them to information and ideas that they can take to their units.

### Promote a Collaborative Environment for Sharing Ideas and Tools That Support Leader Development

Senior officers strongly stated that they need to have flexibility in adapting activities to unit circumstances and their own strengths and experiences. Accordingly, many expressed interest in having a selection of tools and activities that they can apply as needed.

Many of those tools and activities have already been designed by leaders throughout the Army. Unit commanders share their knowledge and resources, but they tend to do it within their own group of friends and professional acquaintances. One role for TRADOC organizations should be to promote wider and more systematic peer-to-peer sharing of best practices. The Center for Army Leadership can support this initiative by compiling and organizing leader development materials from commanders and making them centrally available. Other shared information could be a compilation of vignettes and success stories. There must be some quality control on the resources that are shared, but the burden need not fall entirely on CAL. There are examples from Web-based companies of how to "democratize" quality control by letting the users themselves rate and review resources. These resources and information should be available to the field, but they are likely to be most useful to students in Army schools, particularly those preparing for company, battalion, and brigade command. It may be particularly important that resources are shared with students at earlier stages in their careers, such as AWC and NDU students sharing with pre-command course students, or CGSC students sharing with CCC students.

# Acknowledgments

The authors thank the staff of the Center for Army Leadership, particularly Jon Fallesen, Colonels Mark French and Bruce Reider, Lieutenant Colonels Darin Lewis and Judith Price, and Major Ervin Eddings, for their support and assistance in conducting this research. We also thank the individuals who helped coordinate site visits, including Paul Goodspeed, Rick Travis, Colonels Dorene Hurt, Keith Pickens, and Mark Tillman, Lieutenant Colonels Justin Kidd and Kevin McKenna, Majors Dominick Edwards, David Gordon, Damian Green, and James Kimbrough, and Captain Monica Sneed. Brigadier General Robert Brown and Colonel H. R. McMaster discussed their approaches to leader development as we were beginning this project. Hundreds of Army officers participated in the data collection by completing questionnaires or discussing leader development with teams of Arroyo Center researchers. We are grateful to them for their generous contribution of time. Ricardo Rivera and Lieutenant Colonel Todd Henry provided data for the project. Joanne Ciulla, Donald Forsythe, Kenneth Ruscio, and Thomas Wren of the University of Richmond's Jepson School of Leadership Studies spent an afternoon discussing leader development with the Arroyo researchers. Lieutenant Colonel Carol Redfield and her ROTC faculty at North Carolina State University likewise discussed leader development and reviewed an early version of the questionnaire used to gather data presented in this document. Other individuals who assisted with the development of the questionnaire include Terron Sims, Colonels Raymond Bingham, Jeffrey Holachek, and John McCracken, and Lieutenant Colonel Anthony Reyes. Our RAND colleague Matt Stafford also made useful suggestions to improve the questionnaire. We are especially grateful to our colleagues at RAND, Amy Richardson, Michael Shanley, and Major Glenn Johnson, for their participation in officer interviews, their feedback on briefings and early versions of this draft, and their overall support to the project. We also thank Natalie Ziegler and our late colleague, Annette Parsons, for their assistance in the preparation of this manuscript. Finally, we thank Colonel (Ret.) Jeff McCausland and our RAND colleague Jim Quinlivan, who provided excellent reviews of an earlier draft of this report.

# Abbreviations

| | |
|---|---|
| AAR | after-action review |
| ARI | Army Research Institute |
| ATLDP | Army Training and Leader Development Panel |
| AWC | Army War College |
| CAL | Center for Army Leadership |
| CCC | Captains' Career Course |
| CGSC | Command and General Staff College |
| CO | commanding officer |
| CPX | command post exercise |
| CTC | combat training center |
| DA | Department of the Army |
| DSF | Developmental Support Form |
| FCX | fire coordination exercise |
| FM | field manual |
| FTX | field training exercise |
| HQDA | Headquarters, Department of the Army |
| JODSF | Junior Officer Developmental Support Form |

| | |
|---|---|
| LCX | logistical coordination exercise |
| LDP | leader development program |
| MAPEX | map exercise |
| NCO | noncommissioned officer |
| NDU | National Defense University |
| NTC | National Training Center |
| O/C | observer/controller |
| OER | Officer Evaluation Report |
| OIF | Operation Iraqi Freedom |
| OPD | officer professional development |
| OPTEMPO | operational tempo |
| PAM | pamphlet |
| PDOS | Professional Development of Officers Study |
| PT | physical training |
| ROTC | Reserve Officers' Training Corps |
| S-3 | operations staff officer |
| SOP | standard operating procedure |
| TEWT | tactical exercise without troops |
| TRADOC | U.S. Army Training and Doctrine Command |
| XO | executive officer |

CHAPTER ONE

# Introduction

## Background

The Army, like almost every business and organization, makes experience the cornerstone of its leader development process. Army careers are designed around a series of assignments of increasing scope and responsibility that progressively build experience. The fundamental experience for a soldier is an operational assignment. The early part of an officer's career is spent in an operational unit, with the officer performing the duties of his or her basic branch. Army leaders consistently report that the largest contribution to their development comes from operational assignments.

Soldiers in an operational assignment train constantly to improve collective and individual job performance. Training overwhelmingly focuses on tactics, techniques, and standard operating procedures (SOPs). The business world might refer to this type of training as "management development." The emphasis is on task performance and "the application of proven solutions to known problems, which gives it mainly a training orientation."[1] During deployments, particularly to a hostile environment, after-action reviews (AARs) may supplement or even take the place of some individual or unit-level training as a tool for improving performance. AARs frequently deal with novel solutions and problems, particularly in today's operational environment, but still focus on tactics, techniques, and procedures.

[1] David V. Day, "Leadership Development: A Review in Context," *Leadership Quarterly*, Vol. 11, No. 4, 2000, pp. 581–613.

Leader development includes unit-level training and AARs but is a broader process. The Army's leadership requirements model includes many attributes and competencies besides domain knowledge, such as values, empathy, military bearing, and the ability to create a positive environment, to name a few.[2] Commanders are responsible not only for ensuring that their soldiers are tactically and technically competent, but also for nurturing the leader attributes and competencies of their soldiers. They play a critical role in leader development.

The Army emphasizes the role of unit commanders in developing their subordinates as leaders and provides guidance for how to do so. Field Manual (FM) 7-1, *Battle-Focused Training*, discusses the commander's responsibility for leader development and presents a notional leader development action plan.[3] Department of the Army (DA) Pamphlet (PAM) 350-58, *Leader Development for America's Army*, lists many activities that can be incorporated into operational assignments, such as staff rides, professional development classes, instruction on the history and traditions of the unit, shared experiences, counseling and coaching, and progressive assignments of increasing responsibility.[4] The Chief of Staff of the Army provides professional reading lists for personnel at various leadership levels (e.g., company grade, field grade). These readings could be done individually or incorporated into a unit's leader development classes. The Army also prescribes evaluation and support forms as tools to help develop leaders during their operational assignments; these are designed to provide the basis for verbal counsel-

---

[2] The Army's leadership requirements model can be found in Headquarters, Department of the Army, *Army Leadership: Competent, Confident, and Agile*, FM 6-22, Washington, D.C., October 2006, pp. 2–4. Colonel (Ret.) Jeffrey McCausland, former Dean of Academics at the Army War College, describes these competencies and attributes as being necessary for leadership in an environment of uncertainty. They are, he argues, required increasingly earlier in an officer's career due to advancing technology and a changing strategic environment. See Jeffrey D. McCausland, ed., *Educating Leaders in an Age of Uncertainty: The Future of Military War Colleges*, Carlisle, Pa.: Dickinson College, 2005.

[3] Headquarters, Department of the Army, *Battle Focused Training*, FM 7-1, Washington, D.C., September 2003, pp. A-6 to A-7.

[4] Headquarters, Department of the Army, *Leader Development for America's Army*, Pamphlet 350-58, Washington, D.C., October 1994.

ing and one-on-one interaction between rated leaders and their raters and senior raters.

Most of these activities are not required, and even if they do take place their content and quality may vary greatly. They require time, effort, and commitment by the participants if they are to be truly effective. The current demands placed on the Army by its commitments in Afghanistan and Iraq have made it more difficult to engage in these formal and informal leader development activities. Thus, it might be expected that there would be considerable variance across units in the content, frequency, and perceived quality of leader development programs. But there are few studies that indicate whether units even have leader development programs, and if they do, what the programs consist of and how well they are executed.

## What We Set Out to Do

The Center for Army Leadership (CAL) asked the Arroyo Center to identify feasible and effective leader development programs in operational units. After discussions with CAL, we decided to focus our research on the development of company-grade officers (as opposed to noncommissioned officers [NCOs] or field-grade officers). Operational experience for lieutenants and captains lays the foundation for an entire Army career and thus has a major influence on shaping the next generation of senior leadership. By contrast, development of leaders in higher ranks includes not only operational (including joint) experience, but also more varied experiences such as fellowships, advanced civil schooling, Pentagon tours, and tours within many of the Army's core institutions.

To learn how leader development is being done in operational units, we collected information and opinions from 466 officers. Over 300 of the officers were captains (with a handful of lieutenants); another 84 were (mostly) junior majors; the others were either colonels or lieutenant colonels.[5] The colonels and lieutenant colonels provided insight into, and commentary on, what we had heard from the captains and majors.

We decided the optimal time to talk to officers was when they were in school. By interviewing them in the schools, we met them when they were relatively free from stress and had had sufficient distance and time to reflect on the nature of their entire experience in their previous unit. Had we tried to intercept officers in units, some would have just arrived and might not have fully perceived how leader development was taking place in their unit. On a more practical level, it would have been difficult to interview people in units that were preparing to deploy. In any case, many of the officers in a unit might have been too busy to meet with us.

## How the Report Is Organized

This report describes leader development activities taking place in operational units through the eyes of the officers who participated in those activities. The next chapter describes the officers who participated in this study and the methods used to gather the information. Chapter Three provides an overview of the leader development activities that take place in operational units and what junior leaders say they value most. The critical role of the unit commander as developer of junior officers is the subject of Chapter Four. Chapter Five examines the important processes of counseling, coaching, and mentoring in operational units. These activities are the foundation of the personal interaction that junior officers value highly. Chapter Six presents data reflecting the frequency and content of specific leader development activities, such as reading lists, staff rides, and officer professional

---

5 The written survey did not ask officers to provide their rank, because we assumed that all students at the CCCs would be captains. However, we did observe among the respondents a few lieutenants (fewer than ten). The responses of these officers are included in data characterized as coming from the junior captains, because their presence at the CCCs indicated imminent promotion to captain and they had experience very similar to that of their fellow students.

development (OPD) classes. The concluding chapter summarizes the findings and offers recommendations for how the Army can strengthen unit-level leader development activities.

This report contains several appendixes. The questionnaire that the majors and captains completed is found in Appendix A. Appendix B summarizes responses to the first two items on the questionnaire. The first asked respondents to list the qualities of a specific leader whom they admired and wished to emulate; the second asked respondents what the position of that person was. Appendix C provides essentially verbatim written answers from the majors and captains describing positive and negative leadership lessons they have learned from prior operational experience. Appendix D has an example of a developmental support form that a battalion commander created for her unit. Finally, Appendix E briefly examines what other researchers and organizations have found in their studies of effective leader development programs.

CHAPTER TWO

# Study Participants and Methods

This chapter describes the officers who participated in this study and the methods by which information was collected and analyzed. The pillars of this project were paper questionnaires and face-to-face discussions with 466 Army officers, from lieutenants to colonels. We also reviewed the academic and doctrinal literature and interviewed experts in the field of leader development. Readers who only wish to skim this chapter should look at Table 2.1, which lists the numbers and sources of study participants. Those same readers should also keep in mind when reading subsequent chapters that the officers referred to as "junior captains" were basically platoon leaders during their last operational assignment, while the officers referred to as "majors and senior captains" had more diverse duties in previous assignments, but virtually all had been company commanders, and for many that was their most recent previous assignment.

## Sources and Numbers of Participating Officers

During the summer of 2006, our research team visited five captains' career courses (CCCs), the National Training Center (NTC) and the Command and General Staff College (CGSC) to interview captains and majors and have them complete a questionnaire about leader development activities in their last operational assignment.[1] We met with

[1] Due to limited time and budget, we could not visit all CCCs. But the schools that were chosen represent combat arms (armor, infantry, and engineer), combat support (chemical),

groups of five to ten officers for an hour at a time. During the first half hour of the meetings, the officers completed the questionnaire, which addressed counseling, mentoring, leadership-focused lessons embedded in training exercises, staff rides, OPD or leader development program (LDP) classes, professional reading, self-development planning, and chain-of-command support for leader development (see Appendix A for the complete questionnaire). During the second half hour, we led a discussion with the officers. The discussion addressed the same topics as the questionnaire but allowed for collection of richer information than can easily be conveyed on paper, even with open-ended questions. Much of the value of the discussion was in allowing the officers to hear one another's opinions and experiences. All participants were promised anonymity.

In all, 405 officers completed the written questionnaires and participated in the subsequent discussions. The 405 participating officers can be divided into two subgroups: 1) majors and senior captains and 2) junior captains. The majors and senior captains were CGSC students, observer/controllers (O/Cs) at the NTC, and small group instructors at the CCCs; the junior captains were students at the CCCs. The study included a total of 123 majors and senior captains and 282 junior captains.

Twenty-three of the junior captains and 11 of the majors and senior captains last served in a National Guard or U.S. Army Reserve unit. Their questionnaire responses and discussion comments were very similar to those of the active component officers. All are presented together in this monograph. Separating the responses of reserve component officers from those of active officers would have minimal impact on the tables and figures and no impact at all on the conclusions and recommendations. Nothing in the data gives reason to believe that the leader development activities within reserve component units are significantly different from those in active component units, although a

and combat service support (combined logistics) branches. The original plan called for students in the Military Police CCC at Fort Leonard Wood to participate as well, but ultimately they were excluded from the study due to severe restrictions on their availability.

larger sample of the former would be required before one could draw definitive conclusions.

After the CCC visits (but prior to the NTC or CGSC visits), we reviewed the information we had already collected with colonels and lieutenant colonels ("senior officers" henceforth) who were former battalion and brigade commanders. We met with small groups of senior officers for a selective review of the information that emerged from the questionnaires and discussions with the junior captains at the CCCs.

These meetings served several purposes. First, we asked the senior officers whether the questionnaire findings made sense and comported with their own experiences in command and as junior officers. Second, we collected information from the senior officers about what they did to develop junior officers during their commands. Third, we asked their opinions on how unit-level leader development could be improved (or whether it even needs to be).

The senior officers included 13 students and one faculty member at the National Defense University (NDU), 19 students at the Army War College (AWC), 8 observer/controller team chiefs (all lieutenant colonels) at the NTC, and 20 senior officers currently working in the Washington, D.C., area. The students and faculty member were interviewed at NDU and AWC, and most of the Washington-area senior officers met at RAND's Washington Office. One senior officer was interviewed by telephone, and one in his Pentagon office. Although we spoke to some senior officers individually, we typically met with small groups of three to five. Again, this was done to enable sharing of thoughts and ideas among the officers themselves. Participants from both phases of the study include those listed in Table 2.1.

## Most Recent Operational Experience of Participating Officers

Participating officers last served in a variety of units, but unit representation was uneven for at least two reasons. First, we visited only five CCCs. A branch school tends to have officers who served in a unit of that branch. For example, of the 92 participating students at the Infan-

**Table 2.1**
**Number of Participating Officers by Experience Level and Source**

| Experience Level | Source | Number of Participating Officers |
|---|---|---|
| Junior Captains | Armor CCC | 108 students |
| | Infantry CCC | 92 students |
| | Chemical CCC | 14 students |
| | Engineer CCC | 25 students |
| | Combined Logistics CCC | 43 students |
| | *Subtotal* | *282 junior captains* |
| Majors/ Senior Captains | CGSC | 74 students |
| | NTC | 29 O/Cs |
| | Armor CCC | 10 small group instructors |
| | Infantry CCC | 9 small group instructors |
| | Chemical CCC | 1 small group instructor |
| | *Subtotal* | *123 majors/senior captains* |
| Colonels/ Lieutenant Colonels | NDU | 13 students<br>1 faculty member |
| | AWC | 19 students |
| | NTC | 8 O/Cs |
| | Assigned to DC area | 20 officers |
| | *Subtotal* | *61 colonels/lieutenant colonels* |
| | **Grand Total** | **466 officers** |

try CCC, 73 had last served in an Infantry unit; most of the others had last served in an Armor unit. The global war on terror accounts for a second reason that unit representation was uneven. As of summer 2006, many units were deployed overseas and their officers, therefore, were not in the CCCs. Table 2.2 lists the types of operational units in which the participating majors and captains last served. Table 2.3 lists the parent units in which the officers last served.

**Table 2.2**
**Last Operational Unit, by Branch, for Participating Majors and Captains**

| Branch | Majors/ Senior Captains | Junior Captains |
|---|---|---|
| Infantry | 33 | 101 |
| Armor/Cavalry | 22 | 96 |
| Field Artillery | 12 | 9 |
| Air Defense Artillery | 2 | 3 |
| Aviation | 8 | 7 |
| Special Forces | 0 | 3 |
| Corps of Engineers | 9 | 22 |
| Signal Corps | 3 | 0 |
| Military Police | 3 | 2 |
| Military Intelligence | 3 | 0 |
| Chemical | 1 | 7 |
| Transportation | 2 | 9 |
| Ordnance | 2 | 6 |
| Quartermaster | 0 | 8 |
| Multi-Functional Support | 12 | 14 |
| Other | 10 | 2 |

NOTE: One major and seven junior captains reported multiple unit types. Two majors/senior captains and one junior captain did not report any unit type.

The senior officers who participated in the study commanded a variety of units, as shown in Table 2.4. To assure confidentiality of the senior officers' comments, we do not provide a count of each type of unit commanded, but the table does indicate the range of perspectives and experiences of the senior officers who provided input to this study. Also, Table 2.4 shows that many branches were represented in our discussions with the senior officers, thus broadening our branch cover-

**Table 2.3**
**Last Parent Unit for Participating Majors and Captains**

| Unit | Majors/ Senior Captains | Junior Captains |
|---|---|---|
| 1st Armored Division | 4 | 17 |
| 1st Cavalry Division | 4 | 20 |
| 1st Infantry Division | 15 | 16 |
| 2nd Infantry Division | 8 | 21 |
| 3rd Infantry Division | 14 | 38 |
| 4th Infantry Division | 4 | 5 |
| 10th Mountain Division | 4 | 9 |
| 25th Infantry Division | 5 | 32 |
| 82nd Airborne Division | 10 | 23 |
| 101st Airborne Division (Air Assault) | 7 | 7 |
| Guard or Reserve Unit | 10 | 20 |
| Other | 34 | 77 |

NOTES: Four majors/senior captains and one junior captain did not report any parent unit. Of the 11 majors and 23 junior captains who indicated that their last unit was a National Guard or U.S. Army Reserve unit, one major and three captains indicated that their parent unit was one of the active component divisions. This could happen if reserve component units are assigned to an active unit.

age. Roughly half of the participating senior officers last commanded a combat arms brigade or battalion.

The majors and captains had extensive operational experience. As shown in Table 2.5, the median length of the last operational assignment for the majors and senior captains was 34.5 months; for the junior captains it was 33 months. Junior captains tended to have more recent operational assignments and were significantly more likely to have been deployed during their previous assignment. About a third of the majors and senior captains finished their last operational assignment in 2006, compared with two-thirds of the junior captains. About three-quarters of the majors and senior captains were deployed during their last opera-

**Table 2.4**
**Most Recent Unit Type Commanded by Participating Colonels and Lieutenant Colonels**

| | |
|---|---|
| Infantry Brigade | Infantry Training Battalion |
| Military Police Brigade | Medical Battalion |
| Signal Brigade | Military Intelligence Battalion |
| Training Support Brigade | Multi-Functional Support Battalion |
| Aviation Battalion | Psychological Operations Battalion |
| Aviation Support Battalion | Ranger Battalion |
| Cavalry Squadron | Ranger Training Battalion |
| Contracting Center | Special Forces Battalion |
| Engineer Battalion | Special Warfare Training Battalion |
| Field Artillery Battalion | Transportation Battalion |
| Infantry Battalion | |

**Table 2.5**
**Assignment Lengths and Deployments of Participating Majors and Captains**

| Assignment Length/Deployment Frequency | Majors/ Senior Captains | Junior Captains |
|---|---|---|
| Median length of last operational assignment | 34.5 months | 33 months |
| Completed last operational assignment after January 1, 2006 | 32 percent | 67 percent |
| Completed last operational assignment after July 1, 2005 | 43 percent | 83 percent |
| Deployed at least once during last operational assignment | 75 percent | 89 percent |
| Deployed more than once during last operational assignment | 26 percent | 33 percent |

tional assignment, compared with about 90 percent of junior captains. We did not compile statistics on where the officers were deployed, but many volunteered that information either on the questionnaire or in the discussions. As would be expected, the majority of the deployments were to Iraq or Afghanistan.

Of the 282 junior captains, 247 (88 percent) were platoon leaders in their last operational assignment, and 183 (65 percent) were platoon leaders while deployed. Of the junior captains who were not platoon leaders in their last assignment, 10 were company executive officers (XOs). Of all the 282 junior captains, 173 were company XOs. Thus a total of 257 (91 percent) of the junior captains were either platoon leaders or company XOs or both during their last operational assignment, and 210 (74 percent) were one or the other or both while deployed. The median length of time spent as a platoon leader was 16 months; as company XO, 11 months.

All of the small group instructors (20 total) and captain observer/controllers (19 total) were company commanders in their last operational assignment, and 30 of the 39 (77 percent) were company commanders while deployed. Of the CGSC students and the major observer/controllers, 48 out of 84 (57 percent) were company commanders in their last operational assignment, and 32 out of 84 (38 percent) were company commanders while deployed.

Of the colonels and lieutenant colonels from whom we were able to obtain assignment data, over half had completed command since July 1, 2005, and a quarter had completed command since January 1, 2006. Of the same group, over two-thirds commanded while deployed to either Iraq or Afghanistan.

## Level of Participation

Participation was voluntary. It was essentially an opt-out process for students at the CCCs and for O/Cs at NTC and an opt-in process for the others. At the CCCs, we typically met with part or all of one section of students in a one-hour session. After describing the project and distributing the questionnaire, we told each group that the questionnaire and subsequent discussion were entirely voluntary; they did not have to answer any questions they chose not to answer, nor did they even have to stay for the hour. Out of the hundreds of students at the schools, only two chose not to participate at all. A small percentage chose not to answer certain questions, and, of course, some participated more than others in the post-questionnaire discussions.

The questionnaires and interviews at the CCCs included most if not all students at the schools at the time of our visit. Only certain teams of O/Cs were available during our visit to the NTC. None of the majors or captains there opted out. Overall, it is safe to say that the participants were highly representative of the officers at the institutions we visited.[2]

We obtained a list of former battalion and brigade commanders assigned in the Washington, D.C., area and sent them an email describing the project and inviting them to participate. We stated that we were interested in talking to officers who had commanded an operational battalion or a brigade within the past three years. There were exceptions: A few of the officers had not commanded an operational unit but were invited to participate nonetheless because of their high level of interest.[3]

At the AWC and NDU, a coordinator within each institution notified the students of our study and solicited volunteers. We told the coordinators we wanted to speak with about 20 students at each institution, and a combined total of 33 volunteered and ultimately participated. The senior officers who opted to participate in the study may have different views on leader development from those of their peers who did not volunteer. We do not know.

Although our sample of officers was not random, we believe it provides important insights into the Army's unit leader development programs. As shown above, most of the officers have considerable operational experience, much of it very recent, and they represented an array of branches and command echelons. The comments and questionnaire

---

[2] A convenience sample chooses the individuals who are easiest to reach, or sampling that is done easily. Convenience sampling does not represent the entire population, so it does not support statistically valid analyses.

[3] A total of 114 former battalion and brigade commanders received an email invitation. Twenty officers (17.5 percent) ultimately participated in the study. Of the others, nearly 40 percent did not respond, about 20 percent did not meet our criteria, and about 20 percent could not participate for various reasons or had bad email addresses.

responses from the officers in under-represented branches suggest that our conclusions and recommendations are appropriate for units and unit commanders in all Army branches.[4]

## The Questionnaire

The written questionnaire contained 44 questions (some with multiple parts) that included a combination of closed- and open-ended questions. We asked officers to report the content and frequency of a variety of leader development activities during their last operational assignment.

Development and testing of the questionnaire document was a fairly long process. We compiled a preliminary list of questions based on our meetings with a host of organizations at the Combined Arms Center at Fort Leavenworth, readings of Army documents on leadership and leader development, and interviews with two former brigade commanders well respected for their ability to develop subordinates. Once an initial questionnaire document was created, we field-tested it on groups of captains and majors at the Army Human Resources Command in Alexandria, Virginia, and at the Reserve Officers' Training Corps (ROTC) program at North Carolina State University. We also asked several lieutenant colonels and colonels working in the Washington, D.C., area as Army War College Fellows to review the questionnaire and give us feedback on it, but these officers did not complete the questionnaire themselves.

Based on the feedback we received, a number of questions were added, deleted, or rephrased. One observation that several of our reviewers made was that junior captains often may not know the difference between a map exercise, a situational training exercise, a command post exercise, a tactical exercise without troops, and so forth. We therefore eliminated some questions that distinguished between these different types of training events. Another useful suggestion was that we should ensure that officers understood that our questionnaire was

---

4 But due to the nature of the convenience sample, statistical testing of this hypothesis would not be appropriate.

about more than tactical training; we did so by adding questions at the beginning that asked officers to write about the leadership qualities they most admired in someone they had served under or with. Many subsequent questions referred back to these qualities, for example, "How often did you participate in OPD events focused on leadership qualities and positive leadership lessons such as those you listed at the beginning of the survey?"

By far the most important finding during this testing phase was that officers' responses depended heavily on which commander they had in mind when answering the questions. Our testers would ask us questions like, "Which battalion commander are you asking me about? The first one did a lot and the second one did next to nothing for leader development." Thus even during the developmental phase of the questionnaire we identified a key theme that would come up over and over again once we conducted the actual questionnaires and discussions. The final version of the questionnaire asked officers whether there was a change in command for their unit during their assignment, and if so whether it affected leader development activities. We then instructed respondents to answer the questions in reference to the command climate that most strongly encouraged leadership development in their former unit. Some of the findings presented in this report might therefore be characterized as a best-case or most-optimistic measure of leader development activities in operational units. At the end of the questionnaire, we had an open-ended question that gave respondents a chance to describe the effect that the change in command had on leader development within their unit.

We did not specify that officers had to answer questions with respect to a particular echelon—for example, whether certain events or activities were conducted at division, brigade, battalion, or company level. We considered doing so as the questionnaire was being developed, but several officers who reviewed it felt that lieutenants often might not know at what level certain activities originated. It was clear during the discussions that junior officers' attention is usually focused no more than two levels up, which means that for them, the term "unit" generally refers to a battalion or company.

## The Discussions

After each group of majors and captains completed the questionnaire, they stayed another half hour or so and shared their experiences and opinions about leader development. We guided the discussions with a handful of questions given to almost every group, such as whether their unit had a coherent leader development program, what they did for self-development, and what their unit leaders did to promote leader development. We typically spent two hours with each group of senior officers, although with some groups we only had 60 to 90 minutes. We presented them a series of slides, based on our findings from the majors and captains (primarily the junior captains at the CCCs), and used those as a basis for discussion.

Although we used some slides and questions as guides, the discussions covered a range of issues, and not all issues were raised with (or by) every group. In fact, it was not uncommon for the officers to ask one another questions and engage in a lively discussion among themselves, more than with the RAND team. This was particularly true with the colonels and lieutenant colonels. When that occurred, we simply sat back, took notes, and made sure the conversation stayed on topic.

In such a situation officers are bound to disagree, and none of the findings in this report should be characterized as unanimous opinion or universal experience. Certainly no individual who participated in this process should be assumed to agree with all or any of the findings presented here.

CHAPTER THREE

# Overview of Unit-Level Leader Development

In this chapter we present a summary of what junior officers do for leader development in operational units and what they find most valuable. We found that there is no standard leader development program in operational units, but attempting to impose one would do little good. Unit commanders, whose influence on leader development programs is discussed in the next chapter, decide what to do based on unit circumstances and their own strengths and priorities. Junior leaders say that after operational experience itself, the most effective leader development occurs through interaction with role models, mentors, and peers. This is similar to what corporate executives say is most valuable for their development.

It is important to keep in mind when reviewing results in this report that references to majors, senior captains, and junior captains are descriptions of their rank at the time that they participated in the study and not necessarily their rank when they were last assigned to an operational unit. For example, almost all of the officers referred to as "junior captains" in this document served as lieutenants, and most were platoon leaders, during their last assignment.

## There Is No Standard Leader Development Program

There is no set of activities that could be characterized as a standard or typical unit-level leader development program. Given the array of unique circumstances surrounding different units in an Army at war combined with the unique characteristics of the individuals commanding those units, it is not surprising that we found a wide variance in content, frequency, and perceived quality of leader development activities in operational units. Table 3.1 reflects this variance, which is due to several factors discussed in subsequent chapters—unit roles and missions, location, geographical dispersion, unit readiness, and, most important, the unit commander. The data show how many majors and captains reported that each leader development activity took place in their unit. Note that the numbers reflect percentages of *officers*, not percentages of *units* (as discussed in the previous chapter, participating officers represented a variety of units). Data were compiled from questionnaire responses.

Leader development is not an all-or-nothing program. Some officers might have been in a unit that had a reading list but few OPD classes. Other officers might have gone on a staff ride but were not required to keep a written self-development plan. It appears that most majors and captains did discuss leader development issues with their superiors, probably on an informal basis given the infrequent use of the developmental support forms (discussed in greater detail in Chapter Five). Only about one-third of junior captains went on a staff ride with their last unit, and a similar number were in a unit that had a reading list. Of interest is that these two activities require very different levels of effort. Staff rides are time-consuming to plan and conduct. Publication of a reading list is relatively simple and could probably be done in a matter of hours, far less if one simply uses one of the Army's many sources. Yet neither activity occurs very frequently.

Particularly noteworthy is the fact that, according to DA PAM 623-3, *Evaluation Reporting System*, the review of the self-development form (the Developmental Support Form [DSF], DA Form 67-9-1a) with the rater is mandatory, yet it is the activity that the fewest majors and captains said they did.[1] This is an explicit Army requirement that is not being fulfilled within many units. One easy way to ensure that the DSF gets completed is to require that it be submitted along with the officer's Officer Evaluation Report (OER). However, the OER and the

---

[1] Headquarters, Department of the Army, *Evaluation Reporting System*, Pamphlet 623-3, Washington, D.C., May 2006.

**Table 3.1**
**Percentage of Officers Who Participated in Various Unit-Level Leader Development Activities**

| Activity | Majors/ Senior Captains | Junior Captains |
|---|---|---|
| Commander discussed requirements of upcoming assignments and expectations to a very great extent | 23 | 16 |
| Rater discussed leadership skills at least once a quarter | 60 | 69 |
| Senior rater discussed leadership skills at least once a quarter | 34 | 40 |
| Commander encouraged officers to find a mentor or to mentor others | 52 | 51 |
| Required to have a written self-development plan | 28 | 39 |
| Required to review self-development plan with rater(s) | 21 | 30 |
| Leadership lessons generally embedded in training | 68 | 75 |
| Led or participated in an OPD class focused on leadership at least quarterly | 49 | 56 |
| Required to do professional reading focused on leadership qualities and lessons | 28 | 41 |
| Unit had a reading list | 30 | 33 |
| Unit conducted a staff ride off base | 46 | 34 |

NOTE: Numbers show percentage of *officers*, not percentage of *units*, who participated in these activities.

DSF have different primary purposes. Although counseling related to the OER may be developmental for the junior officer, the OER is fundamentally an assessment tool. The DSF, on the other hand, is a developmental tool. The perception that something may go on an officer's permanent record may inhibit candid discussion of an officer's weaknesses and developmental needs. More importantly, requiring records of the DSF would not address whether the DSF is actually being used developmentally. Of the junior captains who did have a DSF, many reported completing it themselves without help or input from their rater, while others said their chain of command only gave it perfunctory attention.

Our view is that Army policy on DSF enforcement—or lack thereof—does not need to be revised. It is best left to the discretion of the unit commanders, as it basically is today. But since Army Training and Doctrine Command (TRADOC) clearly feels that the support form is a valuable tool, we recommend measures that will systematically introduce junior officers to the form, demonstrate its benefits, and set the expectation that the form will be used in units. This approach is more in keeping with a view of the Army as a profession with a culture characterized by autonomy and self-policing—as distinct, specifically, from a bureaucracy—a distinction made by Don M. Snider in *The Future of the Army Profession.*[2]

The discussions with the majors and captains reflected the same variation across units that the questionnaire responses revealed. One captain, for example, reported having had battalion OPDs weekly, which included learning skills, book reports, and some history. In that same unit, the battalion commander also had breakfast with battery commanders. Another captain described the OPD process as occurring "every once in a while" and doing "vignettes every couple of months," while a third claimed that he had "never seen" an OPD on leadership. There was considerable variability as well in how much planning of leader development took place. One captain reported that his battalion had a formal plan that showed the scheduled activities 30, 60, and 90 days out. The program had a tactical focus but incorporated history and readings such as *Once an Eagle.*[3] Another reported that in his

---

2 Don M. Snider, "The U.S. Army as Profession," in Don M. Snider and Lloyd J. Matthews, eds., *The Future of the Army Profession,* 2nd ed., Boston, Mass.: McGraw-Hill Custom Publishing, 2005, p. 14.

3 A fictional account of an officer who receives a battlefield commission in France in World War I and then becomes a career officer. His career parallels and illustrates the history of the Army through the beginning of the Vietnam War, and embodies numerous lessons in leadership and leader ethics. Anton Myrer, *Once an Eagle*, New York: Harper Collins, 2001.

unit OPDs were always in reaction to some event, rather than being planned in advance.

It is relatively simple to demonstrate variation in the frequency or content of certain leader development activities. It is more difficult to demonstrate variation in quality. We rely on the perceptions of the officers, rather than experimental design, for our assertion that the quality of activities varies. Based on officer comments, there appears to be no standard formula for a high-quality leader development program. Formal counseling, for example, could either be a perfunctory event: "It was our responsibility to turn in our JODSF. [Junior Officer Developmental Support Form; company commanders are supposed to complete the JODSF together with the lieutenants.] It was just a block-check." Or it could receive appropriate time and attention: "My unit had two company commanders who stressed the OER support form, did it twice yearly. They really worked on how to fill it out, and stuck by it for counseling." Another junior captain offered a broader picture of how leader development programs vary in quality: "My unit definitely had a plan. There was a specific program for lieutenants, and we were tested on it. There was required reading with follow-up testing. The battalion commander had guidance for company commanders to follow up on. But the quality at the company level depended on the competence of the company commander."[4]

## Besides Actual Experience, Role Models and Personal Interaction Are Most Valued by Junior Officers

Although the questionnaire focused on ascertaining the frequency and content of various activities, we did attempt to determine what officers regard as the most valuable leader development activities. On the questionnaire, the majors and captains were given a list of twelve activities and were asked to "select three experiences and activities that

[4] A leader development study directed by then-MG Gordon Sullivan during the mid-1980s similarly acknowledged the issue of variation in quality of leader development programs (specifically OPD programs), but offered little detail. See Headquarters, Department of the Army, *Leader Development Study: Final Report*, Washington, D.C., 1988, p. 10.

took place *during the time you were assigned to your former unit* that were *most effective* in developing your leadership qualities and teaching leadership lessons, such as those you listed at the beginning of the survey." Respondents were then asked to rank the top three experiences or activities first, second, and third. The rankings of the activities are presented in Figure 3.1 for majors and senior captains and Figure 3.2 for junior captains.

For both groups there appears to be a threshold between the top six items and the bottom six. The top six items can be lumped into two categories—operational experience and personal interaction. These are clearly more valuable to the junior officers than are the activities that might comprise a formal leader development program, such as staff rides or reading lists. The bottom six also tend to involve more passive listening, reading, and learning.

One explanation for why AARs do not receive a higher ranking is that the questionnaire asked officers what they found most effective for developing leadership skills and teaching leadership lessons, and AARs tend to be task- and mission-oriented. The relatively low ranking thus may indicate that the AARs serve a different purpose, not necessarily that they are done poorly or that they are not valuable for their intended purpose.[5] Also of interest is the fact that all activities were seen as having some value and received some votes. The least effective were the professional forums, which ranked twelfth. Officers tend to see these forums as a source of information, not as a tool for long-term professional development.

We offer an intriguing comparison of how the majors and captains ranked various activities in our study with how 8,000 corporate executives ranked leader development activities in a study conducted by the Center for Corporate Leadership (see Table 3.2).[6] Most striking is the fact that regardless of whether professionals are in the

---

5 Considering this, it is interesting that AARs scored as high as they did.

6 Corporate Leadership Council, *Voice of the Leader: A Qualitative Analysis of Leadership Bench Strength and Development Strategies*, Washington, D.C.: Corporate Executive Board, 2001, as cited by Jack J. Phillips and Lynn Schmidt, *The Leadership Scorecard: ROI for Leaders (Improving Human Performance Series)*, Burlington, Mass.: Elsevier Butterworth-Heinemann, 2004, p. 8.

**Figure 3.1**
**Majors' and Senior Captains' Ranking of Effectiveness of Leader Development Activities**

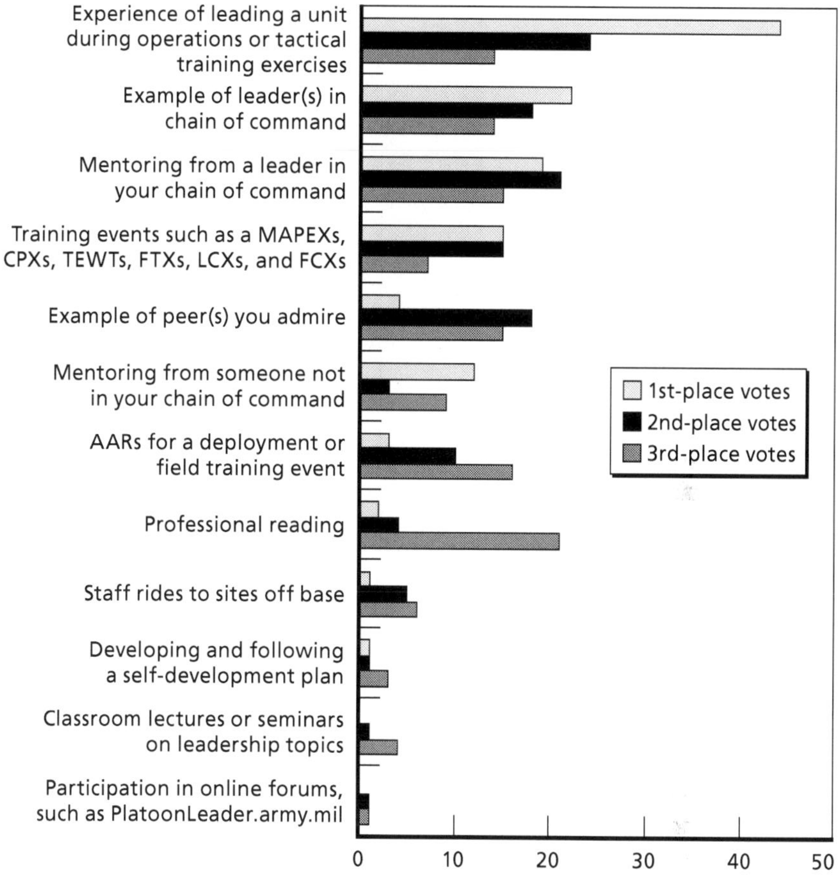


NOTES: MAPEXs = map exercises, CPXs = command post exercises, TEWTs = tactical exercises without troops, FTXs = field training exercises, LCXs = logistical coordination exercises, FCXs = fire coordination exercises, and AARs = after action reviews.

RAND *MG648-3.1*

Army or in the private sector, they highly value personal interaction with coaches, peers, mentors, and counselors. This probably reflects a hard-wired aspect of human nature. Also, having a great amount of decision-making authority is highly valued by the corporate executives. The majors and captains said that operational experience itself

**Figure 3.2**
**Junior Captains' Ranking of Effectiveness of Leader Development Activities**

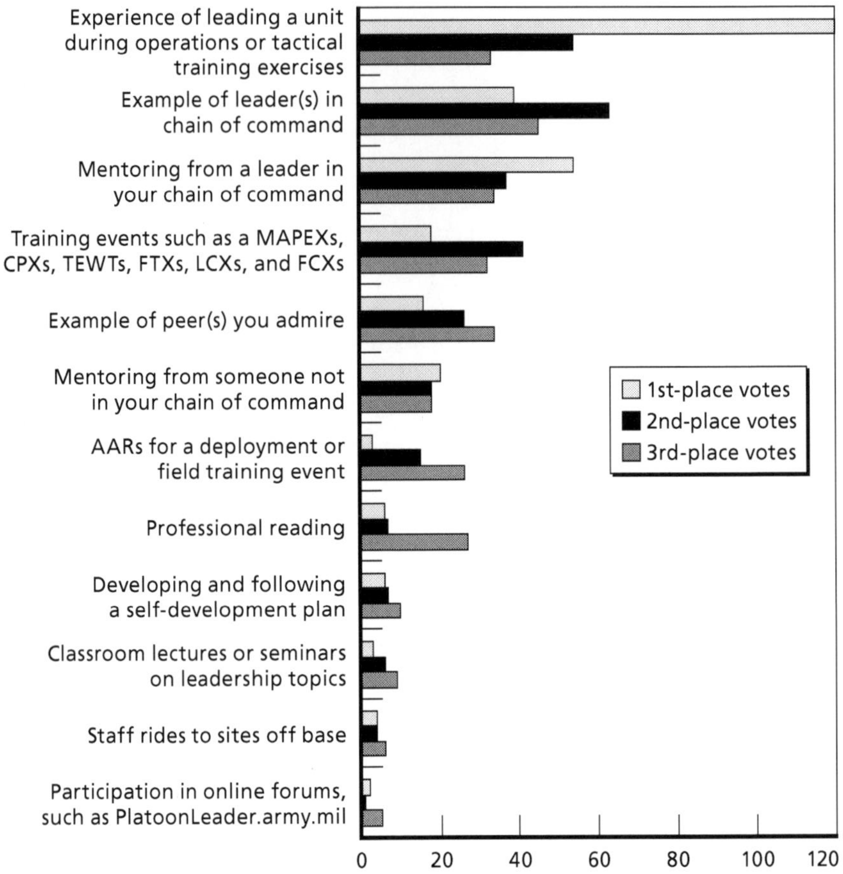


RAND *MG648-3.2*

was most valuable to them. Although the Army questionnaire did not ask about specific characteristics of operational experience, as the corporate survey did, it did ask—elsewhere—about qualities that officers appreciated or admired in a leader. Many officers wrote that they appreciated a leader who delegates authority and allows subordinates to take initiative, suggesting their orientation on the value of broad decision-making authority is similar to that of the corporate executives. Appendix C provides more detail on responses to this question.

**Table 3.2**
**Corporate Executives' Ranking of Importance of Leader Development Methods**

| Rank | Method |
|---|---|
| 1 | Amount of decision-making authority |
| 2 | Creating a leadership development plan |
| 3 | Interacting with peers |
| 4 | Meeting with an executive coach |
| 5 | Meeting with a mentor |
| 6 | Feedback |
| 7 | Turning around a struggling business |
| 8 | People-management skills course |
| 9 | Working in a new functional area |
| 10 | Working in foreign countries |
| 11 | Working in new lines of business |
| 12 | Launching new businesses |
| 13 | Number of direct reports |
| 14 | Quality of direct reports |
| 15 | Off-site seminars in business skills |
| 16 | Technical skills courses |
| 17 | Business skills courses |

SOURCE: Corporate Leadership Council, 2001.

The one seemingly large difference between the two sets of responses is the value ascribed to having a self-development plan. The corporate executives ranked it high, and the majors and captains did not. But, as we discovered, many officers did not keep a self-development plan or did not review it with their rater or senior rater. We have already shown that not all officers participated in each type of leader development activity. It stands to reason that people who did not keep a self-development plan would not select it as one of their three most valuable experiences.

Expanding on this point, we note that if we narrow our focus only to officers who actually participated in certain activities, some of the rankings change. For example and perhaps most notably, among all junior captains, keeping a self-development plan was rated fairly low: ninth overall. But those junior captains who were required to maintain a self-development plan and review it with their rater ranked that as the fifth-most important activity, thus crossing the threshold into the range of things that really matter. This also aligns somewhat more with the corporate executives' rankings. Table 3.3 shows how officers who actually participated in certain activities ranked their value, compared to the ranking from the group as a whole. The findings suggest that the majors and senior captains required to maintain and review a self-development plan did not benefit from that activity as much as the junior captains did, but otherwise the assessments of the two groups are similar. However, we caution that the subset of officers who participated in some of these activities is small—particularly among the majors and senior captains. These findings might best be interpreted as indicators of future areas for additional research, rather than conclusive evaluations of the merit of the different activities.

**Table 3.3**
**Officers' Rankings of Effectiveness of Leader Development Activities If They Participated in Those Activities**

| Activity and Type of Ranking | Majors/ Senior Captains | Junior Captains |
|---|---|---|
| Developing and following a self-development plan | | |
| Overall ranking | 10 | 9 |
| Ranking by officers required to have a written self-development plan | 11 | 7 |
| Ranking by officers also required to review plan with rater(s) | 11 | 5 |
| Professional reading | | |
| Overall ranking | 8 | 8 |
| Ranking by officers required to do professional reading | 5 | 7 |
| Ranking by officers in units with reading list | 5 | 6 |
| Classroom lectures or seminars on leadership topics | | |
| Overall ranking | 11 | 10 |
| Ranking by officers who led a lecture or seminar | 10 | 10 |
| Staff rides | | |
| Overall ranking | 9 | 11 |
| Ranking by officers who went on a staff ride | 5 | 8 |
| Professional forums | | |
| Overall ranking | 12 | 12 |
| Ranking by officers who participated in a professional forum | 11 | 12 |

CHAPTER FOUR

# Commander's Influence on Unit-Level Leader Development Activities

Personal factors are important in all echelons of command. In 2004, the Army War College prepared a study of four Army divisions that had recently returned from Iraq and concluded that "the personal leadership style of the Division Commander remains a unique, significant factor in determining the quality of the command climate."[1] At the battalion level, the unit commander's influence is without question the single most important factor determining the content, frequency, and perceived quality of leader development activities. Unit leaders—especially battalion and squadron commanders—have an enormous influence on the development of junior officers. In our discussions with officers at all levels, a phrase that was frequently used to describe unit-level leader development activities was "personality-driven." The role of the unit commander came up a number of times even in the developmental phase of the questionnaire, and it was the dominant theme of our discussions with officers at all levels.

This chapter describes the influence that commanders have on leader development in a unit, how they function as role models, how they adapt to changing conditions—particularly deployments—and how a change of command affects unit-level leader development activities.

---

1 Walter F. Ulmer, Jr., et al., *Leadership Lessons at Division Command Level—2004,* Carlisle Barracks, Pa.: U.S. Army War College, 2004, p. 9.

## Unit Commanders Affect the Leader Development Activities Within a Unit

Leader development activities vary not only *across* units but also *within* the same unit over time. With a new commander may come a change in the content, frequency, and perceived quality of activities. A common statement from the junior captains would be along the lines of, "I had two battalion commanders. One really focused on developing junior officers. One focused on his in-box, and we never saw him." Or, "I had three company commanders. One took time to develop his lieutenants. One would schedule things but they would always get canceled. One acted like he didn't want to be bothered." Questionnaire responses revealed the same phenomenon. Of the majors and captains whose unit had a change of command, almost nine out of ten said that it affected attitudes or approaches to leader development in their unit.

At the end of the questionnaire, an open-ended question asked officers to briefly describe the effect that the change in command had on leader development activities within their unit. The responses indicated that in some cases the effect was positive: For example, "Our new battalion commander conducted multiple OPDs with his officers weekly. He brings the command together, then the staff and the lieutenants, and [has] them all [together] with their peers." And in some cases the effect was negative: For example, "The previous squadron chain of command was more focused on the formal professional development of officers through OPD. The new one did not utilize OPD as often or as effectively."

Although the battalion commander is the most important individual affecting leader development programs for junior officers, others matter as well. Company commanders, XOs, and operations staff officers (S-3s) were all mentioned in the question asking how a change in command affected leader development in the unit. Staff officers are not in the chain of command, which indicates that not all respondents stayed within the confines of the question, but we cite this to illustrate the effect that individuals have on leader development within a unit.

## Unit Commanders Are Role Models for Subordinates

As Figures 3.1 and 3.2 show, majors and captains ranked "example of leader(s) in chain of command" second only to operational experience itself as the most valuable leader development experience. Junior officers observe *traits*, such as integrity, fairness, and courage, that they wish to cultivate in themselves. Junior officers also observe *actions* that they wish to emulate. For example, a company commander may have one-on-one discussions with his battalion commander about leadership issues, appreciate how much he benefits from those talks, and then decide to do the same with his platoon leaders.

As suggested by the officers who tested the early version of the questionnaire, we added questions at the beginning to focus respondents on leadership qualities and competencies beyond tactics and SOPs. Several of the subsequent questions on the questionnaire referred back to the leader qualities and lessons learned, as a guide to respondents about the types of interactions, activities, and programs we were interested in.

The very first item on the questionnaire asked respondents to describe the leadership qualities of a specific person in the Army who sets an example that they would like to follow. The junior captains emphasized *warrior ethos/professionalism* and *command presence* above other qualities. The responses that we categorized as *warrior ethos/professionalism* included words and phrases such as "courageous," "mentally strong," "determined," and "can-do attitude." The responses that we categorized as *command presence* included words and phrases such as "confident," "assertive," "leads by example," and "leads from the front." Over half of the junior captains listed the first quality, and over 40 percent listed the second. Farther down the list, but still fairly common, were qualities such as *task competence*, *skill in dealing with people*, and *character/integrity*. Each was listed by a third of the junior captains.[2]

---

2 Most officers listed several qualities, which is why percentages for the qualities sum to far more than 100 percent.

Majors and senior captains also had the same top two qualities—*warrior ethos/professionalism* and *command presence*. However, several other qualities received roughly the same emphasis from this group. Those qualities included *task competence*, *empathy*, *character/integrity*, and *communication skills*. No single quality was listed by at least half. Appendix B provides more detail on responses to this question.

The respondents were also asked to indicate the position of the person they had described. By far, majors and senior captains were most likely to write that the leader they wished to emulate was a battalion or squadron commander (43 percent of all majors and senior captains). Junior captains were most likely to list their company commander (35 percent of junior captains), followed by battalion or squadron commander (26 percent). By comparison, just 16 percent of majors and senior captains and only 1 percent of junior captains listed the brigade or regimental commander. This does not mean that the brigade commanders are not worthy of emulation, but that the platoon leaders and company commanders tend to focus their attention on more proximate leaders. Particularly in the current conflicts in Iraq and Afghanistan, there may be considerable physical and operational separation between battalions and their respective brigade headquarters. Young officers serving at widely separated forward operating bases may only have infrequent interaction with their brigade commanders. Among all respondents, NCOs were listed more often than brigade or regimental commanders (8 percent versus 6 percent).[3]

This raises an interesting point regarding the Army's thinking that leaders develop others two echelons down. According to that model, the battalion commanders focus on the platoon leaders, while the brigade commanders focus on the company commanders. Our results would suggest that company commanders as well as platoon leaders look to their battalion or squadron commanders as primary role models.

---

3 Many respondents listed more than one position, and in some cases it was clear that they were describing two distinct leaders (for example, "battalion commander and platoon sergeant") while in other cases it is possible they were describing one leader who had two different positions in the unit (for example, "battalion XO/S-3").

Unfortunately, as we discovered when we were developing the questionnaire, many officers learn what *not* to do by observing what they perceive as poor examples of leadership. Several junior and senior officers who helped us develop the questionnaire suggested that we ask about lessons learned from bad as well as good examples. That was the third item at the beginning of the questionnaire. All usable written responses to that question are provided in Appendix C.

## Commanders Adapt Leader Development Activities to Changing Circumstances of the Unit

The commanders are responsible for taking many factors into consideration when deciding what leader development activities are most important for their units at any given time. In the discussions, officers noted that what a unit did for leader development depended on deployments, the roles and missions of the units, unit location and geographical dispersion, readiness levels, and the amount of individual and team experience. To respond to these factors, a number of senior officers said it would be helpful to have a flexible "tool kit" of leader development ideas. This is discussed more fully in the final chapter. The current chapter explores the key circumstances that commanders must take into account.

Immediately before deployment, higher headquarters may direct certain required activities that could be considered leader development. Also, units frequently ship or transfer responsibility for equipment and vehicles some weeks prior to the actual movement of troops. Unit commanders use this period to emphasize key issues related to the deployment, and leader development activities generally increase, although this is not always the case. Figure 4.1 shows the effect immediately before deployment. The data reflect the responses to the question, "Was there an increase or decrease in the number of leadership development activities and events in the time **immediately preceding** deployment?"

**Figure 4.1**
**Change in Leadership Development Activities Immediately Before Deployment**

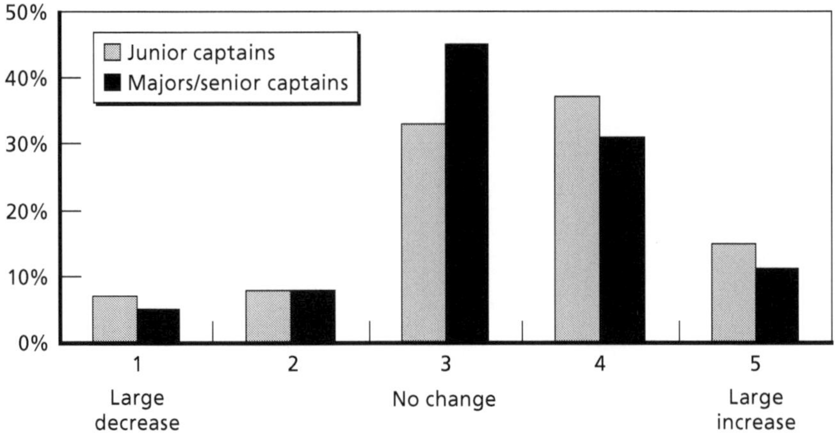


RAND *MG648-4.1*

Fewer than 15 percent of the officers reported some decline. Most reported either no change or an increase, with 15 percent of the junior captains and 11 percent of the majors and senior captains reporting a large increase.[4]

Officers were given an open-ended question that asked how leader development events and activities changed prior to deployment. Of the 97 officers who provided detail about the changes in activities, 41 percent addressed mission-specific tactical and cultural training. For example, one junior captain wrote that they "focused on Iraq . . . culture, religion, enemy tactics. Not so much generic leadership topics." A handful of respondents did note greater emphasis on other types of activities, such as team building.

Once the unit deploys, the press of day-to-day operational demands appears to drive down the level of focused leader development activity. Few officers reported large increases in leader development activities, and many more reported large decreases, relative to what took

[4] For all data regarding effects of deployment, the figures are only for those officers who deployed with their unit. This is a large percentage of all officers, but not the entire group.

place before deployment. Only 9 percent of the junior captains and 3 percent of majors and senior captains reported large increases, while 23 percent of the junior captains and 21 percent of the majors and senior captains reported large decreases. Typical of this trend is the captain who reported that before deployment, his unit had company-level OPDs, but that once deployed to Iraq, "very little" occurred and what did occur was "reactionary." The data in Figure 4.2 show the responses to the question, "Was there an increase or decrease in the number of leadership development activities and events in the time **during** deployment?"

**Figure 4.2**
**Change in Leadership Development Activities During Deployment**

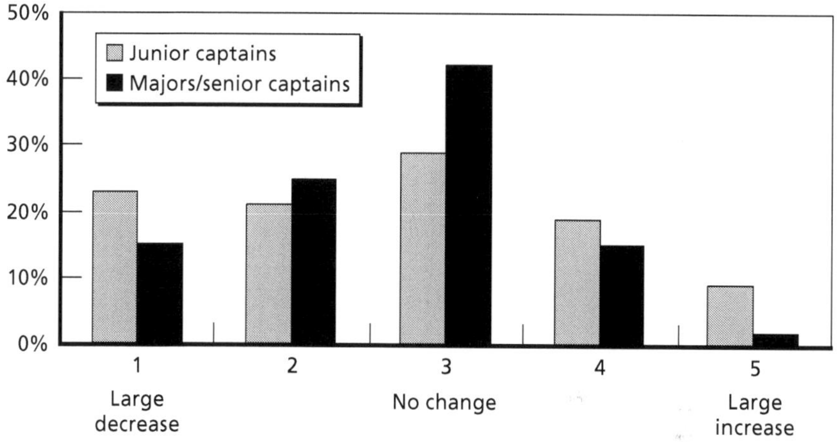


RAND *MG648-4.2*

Activities continue to focus on country-specific, tactical, and mission-focused training during deployment. Officers indicated that the greatest change to their leader development activities was that they became more informal, involving debriefing discussions or one-on-one activities with their superiors. One respondent wrote, "Leadership development was more informal, i.e., just talking to people rather than counseling or OPDs." Also, some respondents noted more attention paid to issues such as stress management, morale, inspiration, and motivation during deployments.

We have less information on some of the other factors affecting leader development activities in units, but we will touch on them briefly. Depending on a unit's roles and missions, the commander may have very close and daily contact with his or her troops or may see them infrequently. Some senior officers told us that they essentially had a captive audience on the forward operating bases and had more opportunities for counseling, OPD classes, and other leader development activities. Others said that because of operational requirements during deployment, they rarely had time for any interaction with junior officers except as it specifically related to their mission for the day.

Related factors are location, geographical dispersion, and unit readiness. Some units had companies and platoons scattered across a huge geographical area in Iraq or Afghanistan. Some units were located in very Spartan, primitive, and dangerous locations. Obviously, in situations such as these, activities such as staff rides or battalion-level OPDs are infeasible.

When units returned from a deployment, some had no vehicles with which to train, so leader development activities became one of the things that filled the void. More commonly, though, commanders of returning units with widespread personnel turnover felt that the most pressing needs were individual and unit training, so leader development activities took a back seat.

One factor deserving special attention is time. Depending on the interaction of all of the other factors, unit commanders may not think they have the time to do leader development or may think their officers do not have time for it. It is their responsibility to set priorities, and we do not question whether those priorities are set properly in any given case. However, although everyone acknowledged that time is in short supply, many officers—at all levels—insisted that there is still enough time to do *something* to develop leaders, even if it is not exactly what the commander, the junior officers, or the Department of the Army ideally would want.

## How Command Influence Affected Responses in the Questionnaire

As discussed in Chapter Two, there were items on the questionnaire addressing the effect of a change in the direct chain of command on leader development within the unit. Almost 90 percent of both groups of respondents reported that there was a change in the chain of command in their last operational unit, and of those who said there was a change, almost 90 percent said the change affected leader development activities in the unit. We instructed respondents to "answer the rest of the questions about your experience *within the command climate that most strongly encouraged* leadership development within your former unit."[5] This was probably easier to do in response to some questions (e.g., yes-no questions such as, "Did your unit have a reading list?") than for others (e.g., frequency questions such as, "How often did you speak to your rater about your leadership qualities?"). To the extent that respondents were able to make this distinction in their minds when completing the questionnaire, their responses present a best-case scenario of how often activities take place and how much they focus on leader development.

---

5 The instructions were also printed in red ink and preceded by the word IMPORTANT.

CHAPTER FIVE

# Counseling, Coaching, and Mentoring

Counseling, coaching, and mentoring are the principal means by which commanders develop leaders. *Army Leadership* (FM 6-22) provides definitions and guidelines for each. Leaders counsel by reviewing with a subordinate the subordinate's demonstrated performance and potential. Coaching tends to focus on skills and tasks. Mentoring occurs through a developmental relationship between a person of greater experience and a person of lesser experience. *Army Leadership* defines different types of counseling: Event counseling covers a specific event or situation; performance counseling reviews a subordinate's duty performance during a specific period of time; professional counseling has a developmental orientation and assists subordinates in identifying and achieving individual goals and goals of their organization.[1]

The value and frequency of these activities was the point of greatest disagreement between the majors and captains on the one hand and the senior officers on the other. Many senior officers insisted that leader development occurs more often than junior officers think, perhaps because it frequently takes place on an informal basis. Several of the senior officers we spoke to said that when they look back on their own careers, they realize now that some of the past conversations they had with their commanders were mentoring sessions, even if they did not realize it at the time. Undoubtedly this happens.[2] We also speculate

1 Headquarters, Department of the Army, October 2006.

2 A similar phenomenon was identified in a study of mentoring at a high-tech manufacturing firm in the Midwest. The researchers plausibly suggested that supervisors know their intentions in each interaction, while the subordinates can only infer them. See Terri A.

that the senior officers were conscious of the example they set while they were brigade and battalion commanders and would consider that a form of mentoring or counseling, whereas junior officers might not. One senior officer even told us that in his experience he would have to remind junior officers about counseling sessions they had forgotten.[3]

But junior officers do recognize that informal interactions are a forum for counseling or mentoring and provide valuable developmental opportunities. Many spoke of commanders who had a weekly or monthly lunch with all the lieutenants, or did physical training (PT) with them, or even did one-to-one activities such as kayaking. Furthermore, even the senior officers frequently said they had not received high-quality counseling more than a few times throughout their own careers. Several senior officers also suggested that much of the interaction with junior officers is really more like coaching, rather than counseling or mentoring, in that it involves immediate feedback on how a certain task or activity is being performed, rather than addressing broader developmental issues.[4]

This chapter reports on what the officers had to say about the type and frequency of the mentoring they received, the frequency of their interactions with their raters and senior raters, the extent of the formal counseling they received, and the support given to their own self-development efforts.

---

Scandura and Chester A. Schriesheim, "Leader-Member Exchange and Supervisor Career Mentoring as Complementary Constructs in Leadership Research," *Academy of Management Journal*, Vol. 37, No. 6., December 1994, pp. 1588–1602.

3 More generally, Dr. Leonard Wong writes that "Generation X captains are clearly more critical of senior officers than captains in the past." Leonard Wong, *Generations Apart: Xers and Boomers in the Officer Corps*, Carlisle Barracks, Pa.: U.S. Army War College, 2000, p. 16.

4 FM 7-1, *Battle Focused Training*, uses the phrase "teach, coach, and mentor" to describe commanders' feedback to subordinates during and after training events. Headquarters, Department of the Army, *Battle Focused Training*, FM 7-1, Washington, D.C., 2003.

## Mentoring

Mentoring appears to occur extensively across units, for majors and captains alike. Responses of the majors, senior captains, and junior captains were so similar on mentoring questions that unless otherwise noted, the figures discussed in this section are for all 405 questionnaire respondents. About 80 percent indicated that they regarded someone in their unit as a mentor. Of those who said they had a mentor, 70 percent said the mentor was in their chain of command. About 60 percent of the respondents said that they had actively sought a mentor in the unit, and about 90 percent of those succeeded in finding one. About half reported that someone in their chain of command encouraged them to find a mentor or encouraged unit leaders to mentor others. This process appears to work, since 89 percent of the majors and senior captains and 80 percent of the junior captains reported mentoring someone else in the unit.

The interviews revealed different definitions and descriptions of mentoring. A number of the senior officers speculated that what a junior officer might say is mentoring is really more like coaching, with a focus on tasks and short-term improvements. Some of the junior captains described mentors in their unit who acted as confidantes—a person to whom the junior officers could reveal uncertainty about how to perform certain tasks or whom they could ask what they feared were simple or dumb questions without being judged. Several senior officers also distinguished between different types of mentors. The type that provides long-term career and professional mentoring often may not be in the same unit as the mentored officer (although they probably would have served together at one time, which is how the relationship is usually established). The mentors may also need to be more than just a couple of years ahead of the people they mentor. Thus it is probably uncommon for a company commander to provide career-oriented mentoring to his or her platoon leaders.[5] Still, the questionnaire responses

5 In the words of one senior officer, captains provide "tactical mentoring" to the lieutenants.

make clear that most officers are able to find someone in their unit who invests personal time and effort into their development as leaders.

This finding is certainly encouraging, and may suggest that a cultural change is under way. When the Army conducted its Professional Development of Officers Study (PDOS) in the mid-1980s, it reported that "too many officers perceive they do not have mentoring leaders."[6] More recently, the Army Training and Leader Development Panel Officer Study found that a majority of officers said they do not have mentors.[7]

## Junior Officers' Discussions with Raters and Senior Raters

On the written questionnaire, the majors and captains were asked, "*During the time you were assigned to your former unit*, did anyone in your chain of command discuss the requirements of upcoming duty assignments and expected standards with you? Please circle the best answer." Over half the officers in both groups are at the high end of the scale in Figure 5.1.

The previous question asked about conversations that may have been essentially duty- and task-oriented. Two other questionnaire items explicitly inquired about conversations addressing the broader topic of leadership skills. One asked, "*During the time you were assigned to your former unit*, how often did you and your **rater** discuss the development of your leadership skills, such as those you listed at the beginning of the survey?" More than a third of the junior captains discussed leadership skills with their rater about once a month or more often; on the other hand, nearly a third did so only once a year, less often, or never. The majors and senior captains gave a similar range of responses, although in general they were less likely to have such a discussion with their rater than the junior captains were (see Figure 5.2).

---

6 Headquarters, Department of the Army, *Professional Development of Officers Study: Final Report for the Chief of Staff, Army, Volume I*, Washington, D.C., 1985, p. 33.

7 U.S. Army Combined Arms Center, *Army Training and Leader Development Panel Officer Study Report to the Army*, Fort Leavenworth, Kan., 2001, p. OS-10

**Figure 5.1**
**Extent to Which Officers Discuss Upcoming Duty Requirements with Someone in Their Chain of Command**

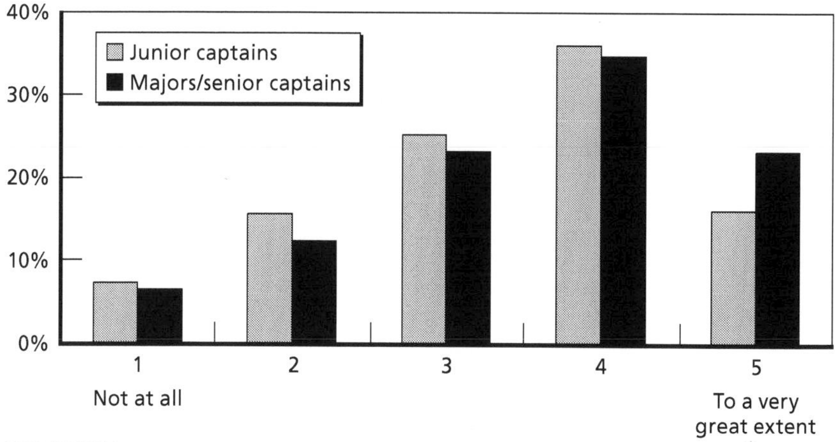


RAND *MG648-5.1*

**Figure 5.2**
**Frequency with Which Officers Discuss Leadership Skills with Their Rater**

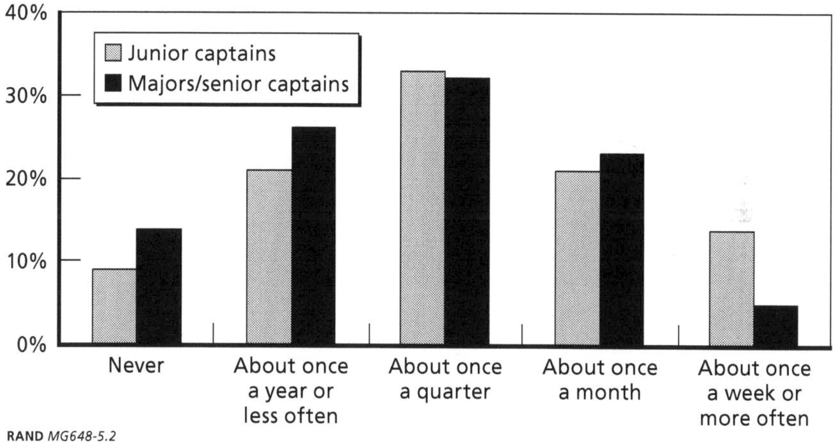


RAND *MG648-5.2*

Conversations with the senior rater about leadership skills were less frequent. Majors and captains were also asked, "*During the time you were assigned to your former unit*, how often did you and your **senior rater** discuss the development of your leadership skills, such as those you listed at the beginning of the survey?" The greatest proportion of both groups of respondents had such a discussion once a year or less often; for majors and senior captains, the second-most-common response was "never." Less than half had such a conversation with their senior rater at least quarterly (see Figure 5.3).

**Figure 5.3**
**Frequency with Which Officers Discuss Leadership Skills with Their Senior Rater**

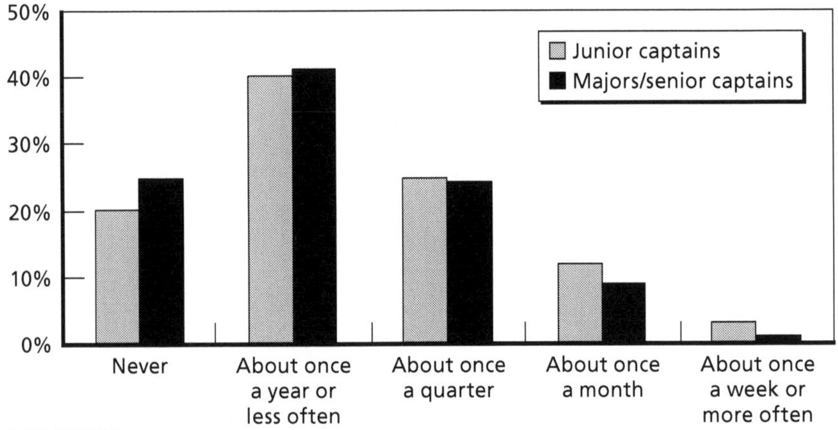


RAND *MG648-5.3*

An interesting story emerges when the responses to the previous two questions are combined, one that reinforces the point that unit commanders have a very important influence on leader development. Data in Figure 5.4 (based only on responses from the junior captains) clearly show that the more often junior captains discussed leadership

**Figure 5.4**
**Frequency of Leadership Skills Discussion with Rater and Senior Rater, for Junior Officers**

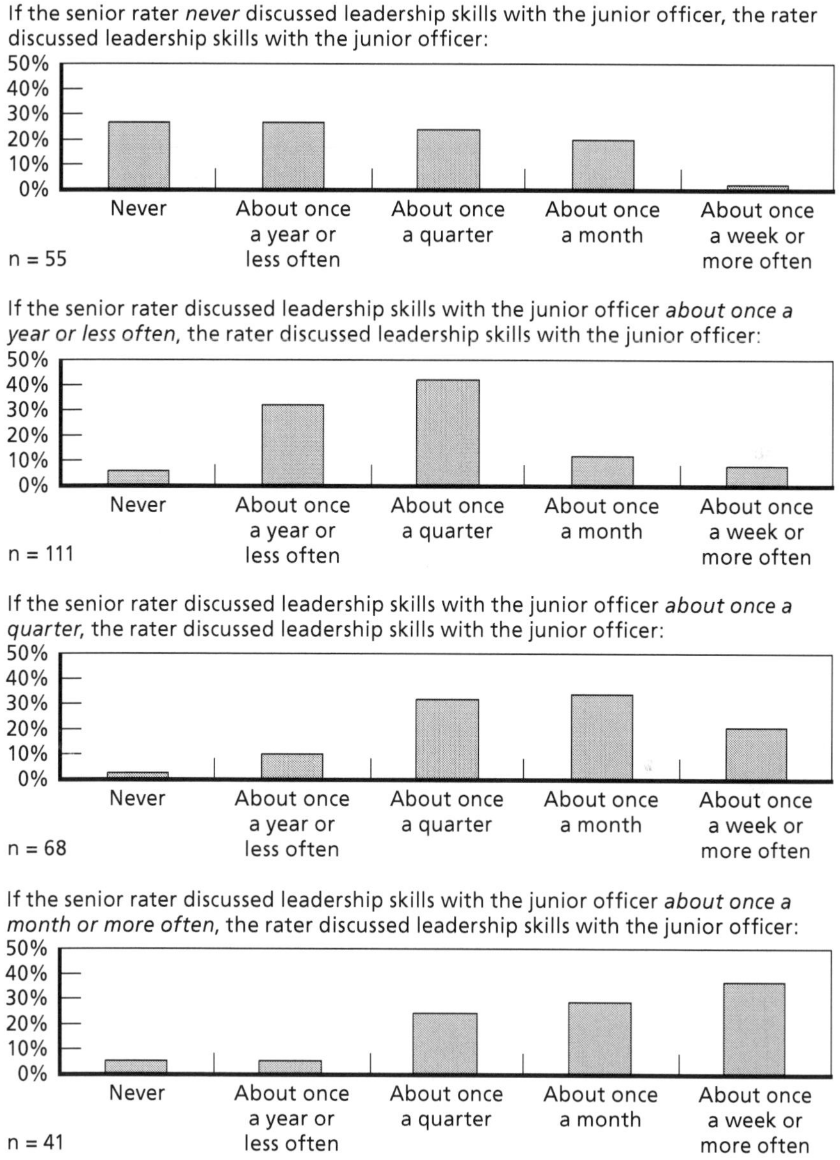


RAND *MG648-5.4*

skills with their senior rater, the more often they did so with their rater as well.[8]

Based on what junior and senior officers alike told us, we offer the following explanation: If the battalion commander (i.e., the senior rater) frequently discusses leadership with platoon leaders, the company commanders (i.e., the raters) understand that this is important and adopt the same behavior. In other words, they "do what the old man says is important." Additionally, it is likely that a battalion commander so engaged with platoon leaders also spends time on leadership issues with his or her company commanders. The company commanders benefit from this interaction and emulate it in their relationships with their immediate subordinates, the platoon leaders.

## Use of Formal Counseling Process

The Army has formal processes for counseling that involve use of the Officer Evaluation Report (DA Form 67-9), the OER support form (DA Form 67-9-1), and the Developmental Support Form (DA Form 67-9-1a). The DSF is supposed to be used as the basis for quarterly developmental counseling of officers in the grades of warrant officer 1, chief warrant officer 2, lieutenant, and captain, as required by DA PAM 623-3, *Evaluation Reporting System*.[9] We also found that many units have their own internal forms that are used in addition to or in lieu of the OER support form or the DSF. Some of the "homemade" forms are career time lines, while others address leadership skills and abilities.

---

8 We present these results only for junior captains for clarity of presentation, but the same phenomenon applies for majors and senior captains. If anything, the data suggest that the trend could be even more pronounced, but so few majors and senior captains report being counseled often by their senior rater that more data should be collected before any conclusions are drawn.

9 The DSF is a written self-development plan that should be initiated within 30 days of the beginning of a rating period, with developmental tasks recorded by the rater and approved and initialed by the senior rater. Follow-up quarterly counseling is to be recorded by the rater on the form.

The questionnaire did not ask about the OER, but during the discussions many majors and captains expressed dissatisfaction with the form and the OER process. Common complaints were that parts of the form are largely meaningless, the OER support form may be completed just days before the OER itself is due, large passages of the OER support form are simply cut and pasted verbatim into the OER itself, and discussions with the rater and senior rater about the OER are frequently short and perfunctory. We did not independently verify this, nor did we systematically raise this specific issue with every group of majors and captains. But frequently they raised the issue themselves.

The questionnaire did ask whether officers were required to maintain a written self-development plan and, if so, whether they were required to review it with their rater or senior rater. Most officers were not required to have a written self-development plan. Only 39 percent of the junior captains reported having such a plan, and among majors and senior captains, the figure drops to 28 percent. The tendency seems to be that the presence of such plans decreases with experience. Nor did much follow-up occur: Of those junior captains who were required to have a written self-development plan, about one-quarter of them never reviewed it with their rater or senior rater.

This finding from the questionnaire was strongly corroborated in our discussions with the majors and captains. A memorable quote from one junior captain was, "Until I became the S-1, I never saw the JODSF [DA Form 67-9-1a was formerly known as the Junior Officer Developmental Support Form]." Another said, "I didn't know your rater was supposed to help you fill out the JODSF until I came here [to the captains' career course]." These statements are characteristic of what we heard time and again: The developmental support forms are not being used as required. As noted above, a number of unit commanders create their own developmental support forms for use by the junior officers in their units. They may or may not require their junior officers to use these forms. Our questions did not specifically ask about the Army's DSF (i.e., the DA Form), so some of the respondents who said they were required to keep a self-development plan presumably

**Table 5.1**
**Self-Development Plans: Frequency of Selected Items**

| Plan Element | Majors/ Senior Captains | Junior Captains |
|---|---|---|
| Plan for sustaining critical skills you had already mastered | 74 percent | 62 percent |
| Plan for training individual proficiency in areas requiring improvement | 94 percent | 95 percent |
| Plan for your future duties or assignments | 77 percent | 70 percent |
| Plan for your future schools or courses | 74 percent | 66 percent |
| Plan for your exposure to new tactical or technical procedures | 57 percent | 63 percent |

NOTE: These are percentages of the 35 majors and senior captains and 109 junior captains who were required to have a written self-development plan.

could have included those whose plan was on a form created by their battalion commander.[10]

The support forms are not submitted to Army Headquarters, and this may help account for their infrequent use. But a number of officers—both junior and senior—opined that the forms are too long, cumbersome, and bureaucratic. The utility of the support forms was one topic of considerable disagreement. Some officers said they simply are not useful. Others said they are. Still others saw them as useful tools for enabling meaningful counseling, but ascribed less value to the forms themselves than to the discussion they spurred.

Turning to the written plans that did exist, they showed some disparity. Almost all plans contained a plan for improving proficiency in areas noted as needing improvement. However, beyond that item, consistency falls off. For junior captains, only two-thirds of the plans had provisions for sustaining critical skills, plans for future schools or

[10] On the other hand, one could argue that had we specifically asked about one or the other Army forms, perhaps we would have spurred the officers' memories and a higher percentage might have responded affirmatively. But nothing in our subsequent discussions with the junior captains led us to believe the support forms are widely used.

courses, or exposure to new tactical or technical procedures. About 70 percent of the plans had a plan for future duties or assignments (see Table 5.1). These questionnaire items were taken from the "Leader Training and Leader Development Action Plan" in Appendix A of FM 7-1, *Battle Focused Training*.[11]

## Supporting Self-Development

Most critical to achieving expertise—as opposed to mere competence—in a field is deliberate practice focused on changing particular aspects of performance. Without it, writes one of the pioneers in the study of expertise, "most professionals reach a stable, average level of performance, and then they maintain this pedestrian level for the rest of their careers."[12] Deliberate practice requires the support and participation of peers, coaches, and mentors, but growth and development will not take place without individual effort.

In our discussions with the groups of majors and captains, the initial response to questions about their self-development effort was usually silence. Even the senior officers had difficulty articulating what a self-development program should comprise. Many officers equate it to reading. This is not surprising, as our review of leader development doctrine found limited guidance on implementing a self-development plan. But the good news is that the desire among junior officers to improve their leadership skills is nearly universal. Many mentioned informal efforts such as role modeling, self-evaluation, and reflection. Having tools to enable understanding of personal strengths and weaknesses could help this process. More consistent counseling, part of an ongoing process of leader development, is another important way to encourage self-assessment and growth.

---

11 Headquarters, Department of the Army, *Battle Focused Training*, FM 7-1, Washington, D.C., 2003.

12 K. Anders Ericsson, "The Influence of Experience and Deliberate Practice on the Development of Superior Expert Performance," in K. Anders Ericsson et al., eds., *The Cambridge Handbook of Expertise and Expert Performance*, Cambridge, UK: Cambridge University Press, 2006, p. 683.

The officers we interviewed expressed keen interest in developing themselves and others as Army leaders. While some aspects of that process such as mentoring seem to be widespread, others are either seldom done or ignored. As is the case with most things at the unit level, commanders play a central role, so any attempts to improve the system must give due regard to that fact. Whatever is done must be done with the goal of assisting the commander rather than increasing his or her requirements. The desire is there. The Army's goal should be to take advantage of it.

CHAPTER SIX

# Specific Elements of Leader Development Programs

The previous chapters addressed key themes in the Army's unit-level leader development programs. This chapter addresses specific activities and events that might compose a formal leader development program.

## Training Exercises

We tried to have some measure of the degree to which leadership lessons are taught via training by asking, "*During the time you were assigned to your former unit*, were leadership lessons (such as those you listed at the beginning of the survey) embedded in training exercises, such as weapons training, FTXs, CPXs, MAPEXs, TEWTs, etc.? Please circle the best answer." The responses show that these training exercises were a modest means of teaching some of the broader leadership issues, but few officers in either group said they were always used to teach leadership (see Figure 6.1).[1]

1 Tables S.2 and 3.1 show that 68 percent of majors and senior captains and 75 percent of junior captains said leadership lessons were generally embedded in training events. To arrive at those percentages, we added the percentages of those who selected 3, 4, or 5 on the scale from 1 (never) to 5 (always).

**Figure 6.1**
**Frequency with Which Leadership Lessons Are Embedded in Training Exercises**

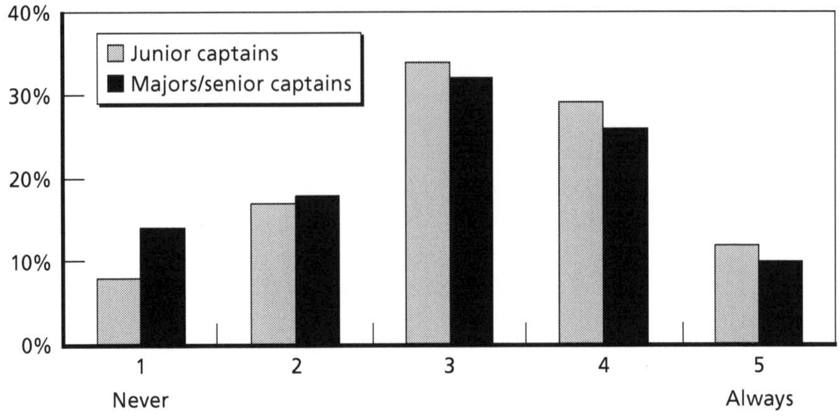


RAND *MG648-6.1*

## Officer Professional Development Classes

When the discussions turned to officer professional development sessions, the consensus among junior officers was that these events tend to focus on unit- or branch-specific skills. Less frequently, they address more general leadership topics. Respondents were given a list of ten broad topics and asked to indicate how often each was the main focus of an OPD or LDP class. Respondents were also asked whether they helped lead or plan a class on each topic. Table 6.1 shows the frequency of various types of OPD or LDP events, depending on the seniority of the respondents. The two groups of respondents reported different frequencies in the coverage of some topics, but the differences are small. Some topics may be more appropriate for platoon leaders than for company commanders, or vice versa. In general, the two groups are in agreement that only unit- and branch-specific skills are the routine subject of OPDs. These topics receive even more intense focus as units near a deployment date.

**Table 6.1**
**Focus of Unit OPD and LDP**

| Frequency | Majors, Senior Captains | Junior Captains |
| --- | --- | --- |
| More than once a quarter | Unit- or branch-specific skills | Unit- or branch-specific skills |
| Two or three times a year | Troop-leading procedures | Troop-leading procedures |
| | How to develop and conduct an effective AAR | Culture, language, and history of a region |
| | Military decision-making process | Ethics |
| | Culture, language, and history of a region | How to develop and conduct an effective AAR |
| Less than once a year, if ever | Ethics | Military decision-making process |
| | How to mentor/coach/ counsel subordinates | Unit/service/military history and traditions |
| | How to build teams | How to build teams |
| | Unit/service/military history and traditions | How to mentor/coach/ counsel subordinates |
| | Cognitive skills, e.g., concentration, visualization, stress management | National security/national defense strategy |
| | National security/national defense strategy | Cognitive skills, e.g., concentration, visualization, stress management |

NOTE: Within the last two frequency categories ("two or three times a year" and "less than once a year, if ever"), activities are listed in order of decreasing frequency.

Table 6.2 shows the percentage of officers who helped lead or plan classes on the various topics. Officers tend to be directly involved in OPD and LDP events, although that involvement varies somewhat based on the topic. For unit- or branch-specific events, 48 percent of the junior captains and 65 percent of the majors and senior captains had either planned or led an OPD or LDP class. Participation in the less frequently presented topics declined correspondingly. For exam-

**Table 6.2**
**Percentage of Officers Who Helped Lead or Plan OPD/LDP Classes on Different Topics**

| Topic | Majors/ Senior Captains | Junior Captains |
|---|---|---|
| Any topic | 81 | 63 |
| Unit- or branch-specific skills | 65 | 48 |
| Troop-leading procedures | 51 | 31 |
| Culture, language, history of a region | 28 | 20 |
| Ethics | 35 | 14 |
| Military decision-making process | 45 | 12 |

ple, only 12 percent of the junior captains helped plan or lead sessions about the military decision-making process. Nearly half of the senior captains and majors helped plan or lead such a session.

## Staff Rides

Staff rides are a type of leader development event that deepens understanding of the recurring fundamentals of military operations. Typically, they focus on a particular battle or a phase of a battle and involve walking the terrain where the battle occurred, discussing the conduct of the battle and the tactics used by the participants. Not all staff rides require physically walking the battlefield, because there are online sites that offer virtual battlefield tours (e.g., Center of Military History). However, a constant characteristic is that these events take substantial planning, preparation, and time to conduct, which tends to limit their frequency.

Questionnaire data show that staff rides off base occur only infrequently. More than half of both respondent groups say they never went on a staff ride during their last assignment. The data appear in Figure 6.2.

**Figure 6.2**
**Frequency of Off-Post Staff Rides**

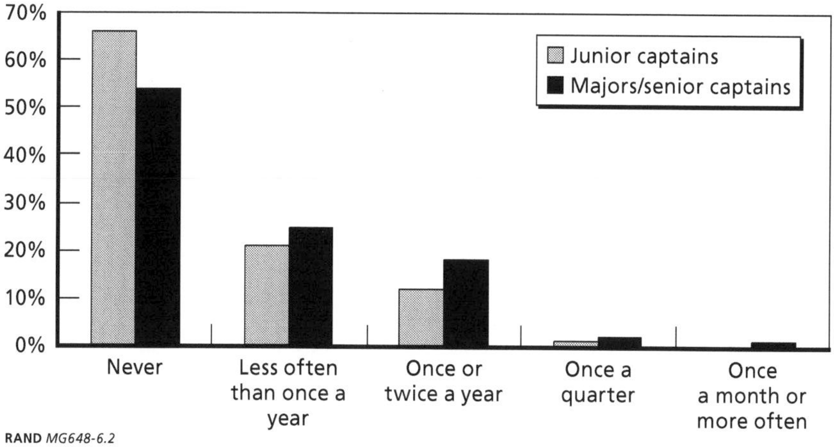


RAND *MG648-6.2*

## Reading Programs

As indicated earlier, most officers do professional development reading. Generally, formal programs are not required. Thirty-five percent of the majors and senior captains and 42 percent of the junior captains report being required to do professional reading. Some units publish a reading list, according to 30 percent of the majors and senior captains and 33 percent of the junior captains. (One-fifth of the respondents did not know whether their unit had such a list.) Regardless of whether it was required, 91 percent of the majors and senior captains and 84 percent of the junior captains read at least one book while assigned to their last unit. The median number of books read during the last assignment was four for the majors and senior captains and three for the junior captains. With the median assignment lasting just under three years, this means officers read about one professionally oriented book a year during an operational assignment.

Comments from the officers indicated that the value of reading books and articles depends on how they are used. A few of the junior captains told us that they were given reading assignments but that there was never any discussion or follow-up. One junior captain told us that his commander gave them a list of books about history and leadership, and they were required to take two with them when the unit deployed. While deployed, the unit did OPDs on the books. "That helped a lot," the junior commander said. Another said when his battalion received a new XO, he implemented a reading program and OPDs focused on leader development. The program was "very good and structured." Reading programs fared reasonably well when officers were asked to select their most valuable leader development activities (see Figures 3.1, 3.2, and Table 3.3). In general, it appears that reading programs can be effective, especially in conjunction with other activities such as OPDs.

Officers also participate in professional online forums such as CompanyCommand or PlatoonLeader, but not in large numbers. Under 40 percent of the participants said they did so (majors and senior captains were slightly more likely to do so). The ones who do use online forums believe they are useful for exchanging job-related information and day-to-day problem solving but less so for longer-term personal and professional development. That is not what the forums were designed for.

## Progressive Assignments

The concept underpinning progressive assignments is that one assignment builds on another. For example, an artillery officer might first be assigned as a forward observer who coordinates and calls for fire support to a maneuver element, then move to an assignment in a firing battery, and then to staff duty in the artillery battalion headquarters, and then back to a battery assignment as a commander.

However, battalion commanders report having limited flexibility in how they assign junior officers. Assignments must be made with unit readiness as the dominant consideration, and individual career progression considerations are addressed within the larger need of the unit, particularly with respect to effectiveness in combat operations. One developmental aspect of the assignments includes pairing junior officers with NCOs who have complementary personalities and greater experience. Commanders may want to distribute the experience of their junior officers and their NCOs throughout the unit by, for example, pairing their least experienced lieutenant with their best platoon sergeant.

CHAPTER SEVEN

# Conclusions and Recommendations

This chapter summarizes our conclusions based on the questionnaire and interview results and presents our recommendations for how the Army might approach improving its leader development programs.

## Conclusions

Army leader development is healthy but could be improved. The questionnaire and discussion findings make it clear that substantial leader development takes place in Army units, but it is equally clear that content, frequency, and perceived quality vary significantly. Some of the variance is due to different circumstances of the units. What they do during routine peacetime training varies from what they do as they prepare for deployment, and that differs from what they do once they deploy. Commanders must also consider their units' missions, readiness level, collective and individual experience, resources, plus a host of other factors, and then must respond and adapt to their units' needs.

But the commanders themselves exert by far the most influence on what takes place in their units, through their personalities, inclinations, effort, and attention to leader development. The commander's influence is so significant that it surfaced even while we were still developing and testing the questionnaire. Unfortunately, some commanders give scant attention to developing their junior officers.

Whether a junior officer serves in a unit with effective leader development activities is a matter of time and chance. Over the course of a three-year assignment, a junior officer's battalion leadership and staff change; some of those battalion commanders, company commanders, and senior NCOs make a perfunctory or half-hearted effort at leader development, others do it very well, and many fall somewhere in between. At some point in the assignment, given the turnover of personnel, the junior officer will likely have a chance to see leader development done well.

The senior officers tended to think that unit leader involvement and the feedback and assessment process are better than was described by the majors and captains, although their disagreement was usually over tone and degree rather than the fundamental accuracy of the observations we collected. The senior officers we spoke to were the ones who volunteered to participate and therefore, because of their commitment to leader development, may have a more favorable view of what takes place in a typical unit than do the more representative group of majors and captains who participated in the study.

The idea of the Army as a learning organization has clearly taken root, and the officer corps embraces the notion that self-development is fundamental to the profession. Not only was this a common theme in the interviews, it is also revealed in the officers' ratings of the most valuable leader development activities. The top items were all related to experience and personal interaction. Learning from the example of others is almost inherently self-developmental, in that it requires reflection and adjustment of one's behaviors and attitudes.

Officers rely on superiors, mentors, and peers not only for self-development, but also for learning how to develop others. Again, this was revealed in both the interviews and the questionnaire results. Colonels and lieutenant colonels said they spoke to previous battalion or brigade commanders when they were deciding how to approach leader development in their own commands. Majors and captains told us how they observed their battalion commanders doing PT with their companies, or ignoring required counseling forms, or staying up late to discuss leadership with young lieutenants after a difficult day in Iraq. The questionnaire data showing how often junior captains discussed leadership skills with their rater and senior rater revealed how leader development activities are influenced by a superior's words and actions. Finally, the questionnaire responses presented in Appendices B and C reveal the range of leadership lessons and qualities that junior officers have learned through observation of others in the Army.

The Army strives to be a learning organization, which is defined in *The Fifth Discipline* as "an organization that is continually expanding its capacity to create its future."[1] That requires mentors, role models, raters, and peers to teach the future generation of leaders. Just as the learning does not end when an officer leaves an Army school and goes back to a unit, neither does the teaching end in the schoolhouse. This phenomenon is implicitly recognized, but perhaps Army culture should more explicitly embrace the teaching aspect of the profession.

## Recommendations

Our recommendations focus on two courses of action that build on the notions of the Army being a learning organization and leaders having a professional responsibility to teach subordinates. Appropriately, they ascribe a critical role to TRADOC organizations. Leader development programs already exist, but arguably their main focus is on development of the individual, whereas the recommendations offered here emphasize development of others. Our recommended strategy is to build on positive processes that already occur every day in the Army, in which people learn from role models, mentors, and peers. We were told by many officers, across all ranks, that people "need to see what right looks like." This approach is the basis for our two key recommendations.

A different approach—one we believe is unlikely to be successful or beneficial—is to impose formal programs, new forms, or reporting requirements on unit commanders. Officers of all grades were emphatic that this would be neither useful nor desirable. Even the idea of a formal mentoring program was unpopular. We learned, as implied earlier, that required counseling is already treated as an administrative burden and

---

1 Peter M. Senge, *The Fifth Discipline*, New York: Doubleday/Currency, 1990, p. 14. Don Snider might argue that despite having the capacity to create its future, the Army nevertheless needs to capture and maintain its renewed professionalism (see Snider and Matthews, 2005). These two views are not at odds but highlight the point that future Army leadership rests on the foundation of ethics, principles, and practices of the profession today.

completed in a perfunctory way by some raters. Commanders would likely resist what they would perceive as additional burdens, whereas the approach we advocate encourages Army leaders to do more of what they are already doing—learning and teaching how to develop others.

### Use TRADOC Institutions to Raise Expectations for Leader Development in Units

Recent history has demonstrated how TRADOC organizations can change Army culture. We heard from many officers that AARs are commonplace—even automatic—in Iraq and Afghanistan today. The senior officers told us that if they did not plan AARs, the junior officers and the NCOs would demand them. This was not the case during the Vietnam War. What happened in the interim? The maneuver combat training centers (CTCs) came into existence, and soldiers that trained there were systematically exposed to the AAR. As a result, what was originally introduced by TRADOC trainers is now an expected and commonplace event in the operational Army. We suggest that the same could happen with leader development by showing officers what right looks like in the schoolhouse, and by exposing them to information and ideas that they can take to their unit.

Counseling is one area in particular where Army schools can demonstrate how leader development should be done. It is an appropriate focus for several reasons. First, counseling is easily adaptable to a school setting. Second and more important, this is an area of relative weakness: Many units do not do counseling according to Army regulations. And third, along the same lines, some officers who do counsel their subordinates do not do it well.

Counseling that takes place within the schoolhouse could address several issues. At the most fundamental level, a schedule should establish when and how often students will meet with instructors or faculty for counseling sessions. This is important to emphasize adherence to a formal process for counseling. Since a large percentage of junior officers are not exposed to the proper use of the DSF, instructors can help students prepare a DSF in anticipation of likely responsibilities during their next operational assignment. They should emphasize to the junior officers that this is something they should expect their rater to do once they get to the unit, and also that this is something they will be expected to do for their subordinates. Students can also prepare and review with their instructor a developmental plan for areas of improvement to focus on while in the Army school itself.

The junior captains told us that what they learned in the schoolhouse about the OER and the DSF focused on administrative requirements, rather than on how they can be used as developmental tools. Instruction should emphasize that there are developmental aspects of the Evaluation Reporting System, and should provide concrete guidance on how to carry out the counseling sessions that accompany the evaluations.

Several senior leaders suggested that since junior officers often do not immediately recognize when counseling or mentoring takes place, it should be more explicit. This is an example that can be set by instructors and faculty by sitting a student down and saying, "Let's talk about your approach to counseling or leadership . . ." Instructors could provide vignettes demonstrating their approaches to leader development, or lead discussions of what the students have observed taking place in units and what they think should take place.

This approach requires that staffing at Army schools be sufficient to allow time for personal attention to students, and that instructor and faculty positions not be treated as being of secondary importance. Preparation of the instructors and faculty should include specific processes and techniques for counseling students, and faculty development may need to be expanded. The relationship between student and faculty member in a school is obviously different from that of an officer and his or her commander, but some of the principles are the same—having regular discussions of developmental needs, setting goals, establishing a plan to accomplish those goals, and showing personal interest in the professional development of others.

The keys to this approach are teaching students through example and establishing expectations for behaviors that embrace the idea of the Army being a teaching as well as a learning organization. If instructors and faculty ensure that students are exposed to formal, personalized, developmental counseling, officers will take their experience and expectations back to the unit and teach others by example.

The goal is to create these experiences and expectations *systematically* within the units, as occurs with AARs today, rather than leaving it to happenstance.

### Promote a Collaborative Environment for Sharing Ideas and Tools That Support Leader Development

Army units represent hundreds of small laboratories for leader development, with people testing new ideas, activities, tools, and approaches. The fact that a number of battalion commanders have independently created their own developmental support forms for junior officers indicates that commanders are not merely being lazy in failing to use the DSF. An example of a battalion commander's support form is provided in Appendix D; its level of detail indicates the amount of thought and effort this commander put into developing her junior officers. Other battalion commanders described career time lines, with personal and professional goals, that their junior officers were required to complete. One colonel had not only made up a reading list for his brigade, he had created slides based on several books and personally led OPD classes in which he presented and discussed the slides. We spoke to other brigade commanders who designed overnight staff rides, reading lists, and OPD classes.

The TRADOC Campaign Plan calls for more distributed learning, local training, and knowledge networks. This applies well to the sharing of ideas and tools for leader development. Army leaders already share knowledge and resources, but they tend to do it within their own group of friends and professional acquaintances.

One role for TRADOC organizations should be to promote wider and more systematic peer-to-peer sharing of best practices. The Center for Army Leadership can support such an initiative by compiling and organizing leader development materials from commanders and making them centrally available. Other organizations besides CAL might also play a role, including the Center for Army Lessons Learned and the Army Heritage and Education Center at the Army War College. Shared information could also include a compilation of vignettes and success stories. The captains and majors said they tend to use online forums not for professional development but to get information. So even though they ranked the forums last in terms of developmental value, they would likely still view them as good sources for information on how to develop others. As one senior officer opined, "The foundation is a collaborative environment. It's even more important today because of generational differences. New lieutenants like sharing information, working in groups."

Quality control is important to making this a successful endeavor. CAL would therefore want to review and parse the resources, stories, and recommendations it receives from the field. In addition, *vox populi* can contribute to quality control. Visitors to Web-based companies such as Amazon or YouTube can review products or videos, giving them both a quantitative and qualitative rating. Other visitors can then search for products or videos sorted in order of rating. Amazon allows others to say how helpful the reviews are, and even lists its top reviewers. This, too, could be adopted as a way to provide some recognition for officers involved in sharing of leader development ideas and materials.

These resources and information should be available to the field, but they are likely to be most useful to students in Army schools, particularly those preparing for company, battalion, and brigade command. On a practical level, online access to these resources may be limited for units in the field, and it is unlikely that a commander will decide in the middle of an assignment to create a leader development program for his or her unit. Students have some time to reflect on past and upcoming assignments, discuss leader development with their peers and colleagues, and share specific tools and ideas. It may be particularly important that resources are shared with students at earlier stages in their careers, such as AWC and NDU students sharing with pre-command course students, or CGSC students sharing with CCC students. A senior officer nicely summarized this idea. He said, "OPTEMPO [operations tempo] is so high that leadership development skills have to be set in the schoolhouse. You need to walk into your job with ideas of how to do leadership development."

Senior officers strongly stated that they need to have flexibility in adapting activities to unit circumstances and their own strengths, experiences, command style, and climate. Accordingly, many expressed interest in having a selection of tools and activities that they can apply as needed. We recognize that the "toolkit approach" cedes to the field some control over the developmental process. In particular, use of the developmental support forms could be seriously undermined if an official Army Web site started providing what is perceived as an alternative to the required form. The tacit message that officers might read into it is, "The required forms and process don't really matter, and I can do whatever I want and call it leader development."

But there is tremendous energy and interest in developing leaders, and what we suggest is a greater role for CAL to guide that energy and reinforce the interest and effort that already exist. By doing so, CAL will, in the words of one senior officer we spoke to, "lower the opportunity cost of creating a leader development program within a unit."

## The Future of Army Leader Development

We conclude with some issues and observations that came to light during the course of this research. Several of the issues raise intriguing questions about how to improve the quality of leader development in the Army.

Many officers, in all grades, mentioned the critical role of NCOs. Perhaps a dozen or so of the captains and majors suggested that NCOs are better at developing subordinates than commissioned officers are. One possible explanation, given by one officer, is that NCOs must carefully document counseling sessions in the event that disciplinary action is required for a soldier. Another suggested that it is because NCOs spend more time with troops. Another simply said that counseling is part of their mentality. Several officers observed that their NCOs were much better at adhering to scheduled formal counseling than the unit's officers were. The nature of the counseling may be different: One major felt that NCOs focus more on skills, which would really have more of a coaching orientation. Two research questions arise: Do NCOs tend not only to counsel better but also to follow Army requirements for counseling better than commissioned officers do? If so, should aspects of their education and training be adopted for officer education and training?

A small number of study participants—11 majors and 23 junior captains—last served in a National Guard or U.S. Army Reserve unit. Their numbers were too small to either draw conclusions about leader development in reserve component units or to affect overall conclusions in this study. The latter point was particularly true because their questionnaire responses were similar to those of the officers who served in active component units. For example, in response to the question of whether the officers felt they had a mentor in their last unit, 78 percent (18 out of 23) of the junior captains from a guard or reserve unit said they did. Of junior captains from an active component unit, 83 percent said they did. Nevertheless, unit circumstances, officer expectations, and culture may all be different enough in National Guard and Army Reserve units to have an effect on leader development that was not discovered during this study.

Officers in all grades struggled to describe self-development. We would ask a group of eight to ten captains or majors what they were doing for self-development, and the typical response was silence. With further questions, many officers would equate self-development to professional reading. Further research could explore why this is so and what could be done to give officers a clearer picture of how they should go about developing themselves professionally.

One particularly challenging question is how to measure and control the quality of leader development activities. We would hear from junior captains about a company commander who used the OER well or a battalion commander who had a "phenomenal OPD program," but it was difficult to pin down what made the OPDs phenomenal or what constituted good use of the OER. We also heard about OPD programs that officers felt were useless, directionless, or boring. Supply and demand can, in a sense, provide a market solution to the question of how to determine quality, provided there is more systematic sharing of resources and tools, i.e., more complete information widely available to "consumers" of leader development ideas. Further research could also make a contribution, perhaps by examining how activities complement and reinforce one another. Research by the Center for Creative

Leadership has found that a serious flaw in many leader development approaches across a variety of organizations is that they are "events-based" rather than systemic.[2] Another line of inquiry could explore the findings in Table 3.3, which showed that how officers rate the effectiveness of various leader development activities depends on whether they participated in those activities. The phenomenon did not apply equally to all activities for the majors/senior captains and the junior captains. Further study could validate this result with larger sample sizes and, assuming validation, look at quality and content of the activities as explanatory variables for why this occurs.

A few senior officers expressed concern that changes in how units are manned, via modularity and lifecycle manning, could have undesirable consequences for leader development. Some branches may no longer have a battalion commander of that branch to mentor and develop their junior officers within modular brigades. With lifecycle manning, junior officers may see less variety in the officers they serve under and with. On the one hand, this could be beneficial in a couple of ways—it will give unit commanders more time to get to know their subordinates, and it may give unit commanders greater incentive to invest time in developing their subordinates. But on the other hand, it could be detrimental for some officers who do not have the opportunity to serve under a commander who is committed to developing his or her subordinates. Thus, there will be a higher premium on ensuring that all commanders execute leader development programs to an acceptable standard. As modularity and lifecycle manning are implemented, further studies of leader development activities within the units may answer the question of whether the senior leaders' concerns are being realized.

Other senior officers mentioned generational issues and observed that things that worked for today's colonels and lieutenant colonels may not work for today's captains and lieutenants. Specifically, many

---

2 Ellen Van Velsor and Cynthia D. McCauley, "Our View of Leadership Development," in Ellen Van Velsor and Cynthia D. McCauley, eds., *The Center for Creative Leadership Handbook of Leadership Development,* 2nd ed., San Francisco, Calif.: Jossey-Bass Publishers, 2004, p. 18.

of the senior officers speculated that junior officers want more interaction and feedback from superiors than did previous generations. As today's junior leaders progress through their careers, they may naturally be more inclined to provide feedback, assessment, and counseling, and they may also be more inclined to communicate via peer networks about how to do so. The Army's Strategic Studies Institute has examined differences in attitudes and professional goals between Generation Xers and Baby Boomers.[3] Longitudinal studies may help reveal the degree to which the views of junior officers truly represent a generational shift as opposed to simply a different perspective from the battalion and brigade commanders who are responsible for developing their subordinates.

Many officers, in all grades, questioned whether retention challenges are tied to dissatisfaction with leader development.[4] In this context, one senior officer told us that when he became a battalion commander, his brigade commander told him, "You're still in the Army because of the first or second battalion commander you had." We never raised the issue of retention in any of the discussions, but captains and colonels alike raised it themselves countless times. Previous research by CAL, examining data collected during the Army Training and Leader Development Panel study and more recently in a 2006 online survey, has corroborated the officers' belief that leader development is related to retention issues. Additional research could examine the temporal relationship between officer separation and dissatisfaction with leader development. Is dissatisfaction something that festers over years, or is it affected mainly by experiences in the last assignment? To what extent are the experiences connected with leader development? How soon in their careers do some officers begin to develop an unfavorable

---

3 Wong, 2000.

4 Note that this is a somewhat different, although related, formulation of the relationship between retention and leadership found in analyses such as that of Mark Lewis, "Army Transformation and the Junior Officer Exodus," *Armed Forces and Society*, Volume 31, Number 1, Fall 2004. Lewis argues that not only do departing officers complain of the quality of their leaders, but that those leaving are more likely to be of higher quality, thus creating a vicious circle of attrition.

opinion of the quality of leader development, and to what extent is it reversible?

The last issue we raise will be addressed over time through simple observation, not research and studies. A couple of senior officers, in two different sessions, offered an intriguing prediction about what the current generation of junior officers will mean to the future of the Army. One senior officer noted that it was the young combat veterans of the Vietnam War who rebuilt the Army in the late 1970s and early 1980s. In that same vein, another senior officer in a different meeting said, "We [today's senior officers] may not be the ones who end up transforming the Army." Today's junior officers—perhaps the ones who will truly transform the Army—are learning many lessons through experience, including how to develop the officers of tomorrow.

APPENDIX A

# Junior Officer Questionnaire

Of the 466 officers who participated in this study, the 405 in the grade of major or below completed a written questionnaire. Two versions of the questionnaire were used, depending on the seniority of the officers, but the only difference was in the list of positions that officers had in their previous assignment for Question 12. The version here was used for junior captains at the CCCs.

This survey:

- Gives you a chance to affect unit-level activities that will make you, your peers, your commanders, and your NCOs better leaders.
- Is part of a study sponsored by TRADOC and supports the work of the Center for Army Leadership.
- Asks about leadership development activities conducted in the last operational unit to which you were assigned.
- Is ***not*** an audit of your activities or those of your chain of command.
- Collects information that will be kept confidential. We will not associate your responses with you or with the particular unit to which you were assigned.

**Section I: Definition of Leadership Development**

1. Think of a specific person in the Army who is a good and effective leader—someone who sets an example that you would like to follow. This could be someone in your chain of command, a peer, or a subordinate. Please list that person's qualities as a leader that you *most admire* and would *like to emulate*.

2. If you wish, please tell us what position this person held when you knew or served with him or her (e.g. battalion commander, fellow platoon leader, company first sergeant, etc.).

3. Please briefly describe lessons you have learned from good or bad examples of leadership that you have experienced in the Army.

<u>IMPORTANT:</u> We will ask you about activities and programs during your last operational assignment that taught positive leadership lessons and gave you an opportunity to develop the leadership qualities you admire. This is our working definition of leadership development. In particular, we want to focus on leadership development that does more than build technical and tactical proficiency.

**Section II: Background information**

The following survey questions ask about your leadership development experiences *in the TOE unit you were most recently assigned to*. We will term that your "former unit." For most respondents, that unit would be:

- A battalion or squadron, or
- A separate company within a brigade, division, or corps– such as a headquarters company, a support company, a signal company, etc.

Unless otherwise specified, the following survey questions refer to all the time you spent while assigned to that TOE unit, including all positions you held in that unit.

4. From the lists below, please select the branch, echelon, and parent unit that best describes your former unit. For example, if you were assigned to the $1^{st}$ Battalion, $22^{nd}$ Infantry Regiment of the $4^{th}$ Infantry Division, select Infantry, Battalion, and $4^{th}$ Infantry Division from the lists below:

| Branch of Unit | TOE Unit Echelon | Parent Unit | Guard or Reserve Unit? |
|---|---|---|---|
| a. Infantry<br>b. Armor<br>c. Field Artillery<br>d. Air Defense Artillery<br>e. Aviation<br>f. Special Forces<br>g. Corps of Engineers<br>h. Signal Corps<br>i. Military Police<br>j. Military Intelligence<br>k. Chemical<br>l. Adjutant General<br>m. Finance<br>n. Transportation<br>o. Ordnance<br>p. Quartermaster<br>q. Multi-functional support<br>r. Other (please specify) | a. Battalion or squadron<br><br>b. HQ company of a brigade or brigade-equivalent<br><br>c. HQ company of a division or corps<br><br>d. Company, battery, troop as a separate asset of a brigade, division, or corps<br><br>e. Other (please describe): | a. $1^{st}$ Armor Div<br>b. $1^{st}$ Cavalry Div<br>c. $1^{st}$ Infantry Div<br>d. $2^{nd}$ Infantry Div<br>e. $3^{rd}$ Infantry Div<br>f. $4^{th}$ Infantry Div<br>g. $10^{th}$ Mountain Div<br>h. $25^{th}$ Infantry Div<br>i. $82^{nd}$ Airborne Div<br>j. $101^{st}$ Airborne Div<br>k. Other (please specify) | a. Yes<br>b. No |

5. What was your branch/functional area while you were assigned to your former unit?

6. Was your former unit part of a Stryker Brigade Combat Team (SBCT)?
a. Yes
b. No

7. In what month and year did your assignment to your former unit begin?
Month____________
Year______________

8. In what month and year did your assignment to your former unit end?
Month____________
Year______________

9. *During the time you were assigned to your former unit*, how many times did you deploy with all or part of your unit (e.g. OIF, OEF, some other operational deployment)?

10. *During the time you were assigned to your former unit*, how many months were you on operational deployment(s) with all or part of your unit?

11. *During the time you were assigned to your former unit*, how many times did you have a training rotation with all or part of your unit at a combat training center (CTC)?

12. In the table below, please indicate how many months you were assigned to each position that you held in your former unit. Please indicate whether you held this position during an operational deployment:

| **Position** | **Enter total number of months you were assigned to this type position** | **Did you hold this position during an operational deployment (circle one)** |
|---|---|---|
| Platoon Leader | | Yes No |
| Company XO | | Yes No |
| Company Commander | | Yes No |
| Battalion Staff Officer | | Yes No |
| Brigade or higher Staff Officer | | Yes No |
| Other (list) | | Yes No |

A reminder: Unless otherwise specified, the following questions ask about *all of your experiences* while you were assigned to your former unit, *including all of the positions you listed above*.

**Section III: Chain of Command Support of Leader Development**

13. *During the time you were assigned to your former unit*, was there a change in your direct chain of command?
a. Yes
b. No

14. Did the change result in a noticeable change in command climate regarding attitudes or approaches to leadership development?
a. Yes
b. No
c. not applicable—no change in your direct chain of command during the time you were assigned to your former unit

IMPORTANT: If you answered "Yes" to the previous question, please answer the rest of questions about your experiences *within the command climate that most strongly encouraged* leadership development in your former unit. There will be a question at the end of this survey that asks you how the change in command climate affected leadership development within your former unit.

15. *During the time you were assigned to your former unit*, did anyone in your chain of command discuss the requirements of upcoming duty assignments and expected standards with you? Please circle the best answer:

| 5 | 4 | 3 | 2 | 1 |
|---|---|---|---|---|
| To a very great extent | | | | Not at all |

16. *When you were in your last leadership position in your former unit*, how often did you and your **rater** discuss the development of your leadership skills (focusing on leadership qualities and positive leadership lessons, such as those you listed at the beginning of the survey)?
a. About once a week or more often
b. About once a month
c. About once a quarter
d. About once a year or less often
e. Never

17. *When you were in your last leadership position in your former unit*, how often did you and your **senior rater** discuss the development of your leadership skills (focusing on leadership qualities and positive leadership lessons, such as those you listed at the beginning of the survey)?
a. About once a week or more often
b. About once a month
c. About once a quarter
d. About once a year or less often
e. Never

**Section IV: Mentoring**

FM 6-22 (*Army Leadership*) defines mentoring as a "voluntary developmental relationship that exists between a person of greater experience and a person of lesser experience that is characterized by mutual trust and respect." The mentor does not have to be of greater rank. For example a senior NCO could mentor a lieutenant.

18. *During the time you were assigned to your former unit*, did your chain of command encourage you to find a mentor or encourage leaders to mentor others?
a. Yes
b. No

19. *During the time you were assigned to your former unit*, did you actively try to find someone within your unit who would mentor you?
a. Yes
b. No

20. *During the time you were assigned to your former unit*, was there someone in your unit whom you would consider a mentor to you?
a. Yes
b. No

21. Was the mentor in your chain of command?
a. Yes
b. No
c. not applicable—there was nobody in your former unit whom you would consider a mentor

22. *During the time you were assigned to your former unit*, did you feel that you were capable of reaching out to others who might seek you as a mentor?
a. Yes
b. No

23. *During the time you were assigned to your former unit*, was there someone in your unit whom you mentored?
a. Yes
b. No

**Section V: Leadership Development Activities**

24. *During the time you were assigned to your former unit*, were leadership lessons (such as those you listed at the beginning of the survey) embedded in training exercises, such as weapons training, FTXs, CPXs, MAPEXs, TEWTs, etc? Please circle the best answer:

| 5 | 4 | 3 | 2 | 1 |
|---|---|---|---|---|
| Always | | | | Never |

25. *During the time you were assigned to your former unit*, how often did your unit conduct a staff ride to a site off base, such as a Civil War battlefield?
a. Once a month or more often
b. Once a quarter
c. Once or twice a year
d. Less often than once a year
e. Never

26. *During the time you were assigned to your former unit*, how often did you **lead** OPD or LPD events focused on leadership qualities and positive leadership lessons, such as those you listed at the beginning of the survey?
a. Once a month or more often
b. Once a quarter
c. Once or twice a year
d. Less often than once a year
e. Never

27. *During the time you were assigned to your former unit*, how often did you **participate in** (but not lead) OPD or LPD events focused on leadership qualities and positive leadership lessons, such as those you listed at the beginning of the survey?
a. Once a month or more often
b. Once a quarter
c. Once or twice a year
d. Less often than once a year
e. Never

28. In the table below, please place a check in the column that indicates how often each topic was the main focus of one or more professional development (OPD or LPD) classroom lectures or seminars that took place *during the time you were with your former unit*. Please indicate whether you helped lead or plan a class on that topic:

| **Topic** | Once a month or more often | Once a quarter | Once or twice a year | Less often than once a year | Never | You helped **lead** a class on this topic (circle one) | You helped **plan** a class on this topic (circle one) |
|---|---|---|---|---|---|---|---|
| Ethics | | | | | | Yes No | Yes No |
| National security/national defense strategy | | | | | | Yes No | Yes No |
| Unit/service/military history & traditions | | | | | | Yes No | Yes No |
| Troop-leading procedures | | | | | | Yes No | Yes No |
| Military decision-making process | | | | | | Yes No | Yes No |
| Unit- or branch-specific skills | | | | | | Yes No | Yes No |
| Cognitive skills, e.g., concentration, visualization, stress management | | | | | | Yes No | Yes No |
| How to develop and conduct an effective after action review (AAR) | | | | | | Yes No | Yes No |
| How to mentor/coach/counsel subordinates | | | | | | Yes No | Yes No |
| How to build teams | | | | | | Yes No | Yes No |
| Culture, language, history of a country or region to prepare for a deployment | | | | | | Yes No | Yes No |
| Other (please specify) | | | | | | Yes No | Yes No |

**Section VI: Professional Reading and Forums**

29. *During the time you were assigned to your former unit*, were you formally required to do professional reading focused on leadership qualities and positive leadership lessons, such as those you listed at the beginning of the survey?
a. Yes
b. No

30. *During the time you were assigned to your former unit*, approximately how many books and articles did you read that could be considered professional reading focused on leadership development?
Number of books: __________
Number of articles, including online articles: __________

31. *During the time you were assigned to your former unit*, did your unit or parent unit have a reading list?
a. Yes
b. No
c. Don't know

32. *During the time you were assigned to your former unit*, did you participate in Army leader development forums (e.g., BCKS, CompanyCommand.com, PlatoonLeader.army.mil, etc.)?
a. Yes
b. No

**Section VII: Self-Development Plan and Activities**

33. *During the time you were assigned to your former unit*, were you required to have a written self-development plan?
a. Yes
b. No

If you answered yes to the previous question, please answer the next two questions. If not, please skip to the next section.

34. *During the time you were assigned to your former unit*, did you review your written self-development plan with your rater or senior rater?
a. Yes
b. No

35. In the table below, please indicate whether each item was part of your written self-development plan *during the time you were assigned to your former unit*.

| Leader Development and Training Items | Your written self-development plan included this item (circle one) | |
|---|---|---|
| Plan for sustaining critical skills you had already mastered | Yes | No |
| Plan for training individual proficiency in areas requiring improvement | Yes | No |
| Plan for your future duties or assignments | Yes | No |
| Plan for your future schools or courses | Yes | No |
| Plan for your exposure to new tactical or technical procedures | Yes | No |

**Section VIII: Effects of Deployment on Leadership Development Activities in Units**

If your unit did not deploy during your assignment, please skip to the next section.

36. Was there an increase or decrease in the number of leadership development activities and events in the time *immediately preceding* deployment? Please circle the best answer:

| 5 | 4 | 3 | 2 | 1 |
|---|---|---|---|---|
| Large increase | | No change | | Large decrease |

37. Was there an increase or decrease in the number of leadership development activities and events in the time *during* deployment? Please circle the best answer:

| 5 | 4 | 3 | 2 | 1 |
|---|---|---|---|---|
| Large increase | | No change | | Large decrease |

38. Did the types of leadership development activities and events change in the time *immediately preceding* deployment? If yes, in what ways?

39. Were the types of leadership development activities and events *during* deployment different from those before or after deployment? If yes, in what ways?

**Section IX: Your Opinions on Leadership Development Activities in Units**

40. In the table below, please select *three* experiences and activities that took place *during the time you were assigned to your former unit* that were *most effective* in developing your leadership qualities and teaching leadership lessons, such as those you listed at the beginning of the survey. Rank the top three 1, 2, and 3, respectively.

Your responses should look something like this ⇒

| |
|---|
| |
| 3 |
| |
| |
| 1 |
| 2 |
| |
| |

or this ⇒

| |
|---|
| 1 |
| |
| |
| 3 |
| |
| |
| 2 |
| |

etc.

| **Activity or Experience** | **Select the three most effective leadership development experiences *in your former unit* (Rank them 1 through 3)** |
|---|---|
| Classroom lectures or seminars on leadership topics | |
| Staff rides to sites off base | |
| Professional reading | |
| Mentoring from a leader in your chain of command | |
| Mentoring from someone not in your chain of command | |
| Example of leader(s) in chain of command | |
| Example of peer(s) you admire | |
| Participation in on-line forums, such as PlatoonLeader.army.mil | |
| Developing and following a self-development plan | |
| AARs for a deployment or field training event | |
| Training events such as a MAPEX, CPX, TEWT, FTX, LCX, FCX | |
| Experience of leading a unit during operations or tactical training exercises, including CTCs | |

Please briefly explain why you chose those three:

41. *During the time you were assigned to your former unit*, were there other activities, events, or programs focused on leadership development that were not mentioned in this survey (in other words, did we miss anything)? If yes, please describe briefly:

42. Previously, we asked about a change in command climate as the result of a new battalion or brigade commander. If there was a change in command climate that affected leadership development during the time you were assigned to your former unit, please briefly describe what effect the change in command climate had on leadership development activities within your unit.

43. *During the time you were assigned to your former unit*, what was the biggest obstacle to having better leadership development activities, events, and programs?

44. Is there anything else you think the Center for Army Leadership should know about leadership development activities, events, and programs in operational units?

45. If we have some follow-up questions, would you be willing to answer them by phone or e-mail? If so, please provide contact information where we could reach you during the next few months.

APPENDIX B

# Leadership Qualities That Junior Officers Most Admire and Wish to Emulate

To focus the officers' thinking on issues broader than tactics and training, there were three items at the beginning of the questionnaire that asked respondents to 1) list the characteristics of a leader they admired; 2) state what position that person held; and 3) describe lessons learned from good or bad examples of leadership that they have experienced in the Army. This appendix summarizes the written answers to the first two questions.

The responses to the first question were wide-ranging. We attempted to group them into a reasonable number of categories for ease of understanding. We drew from two sources to create the larger categories. The first is the Army's Leadership Requirements Model, which can be found in FM 6-22, *Army Leadership*. It lists 20 leader attributes and competencies, together with descriptions of behaviors that exemplify them. The second source is a list of leadership attributes identified by a leading scholar of organizational leadership, John Gardner, who compiled a list of 14 leader attributes that he says are commonly found in successful leaders in North American organizations.[1] We ended up with a list of 11 leadership qualities that reasonably summarize the diverse responses given by the majors and captains on the questionnaire.

Table B.1 shows how often different leadership qualities were listed, together with descriptive words and phrases that we categorized

---

1 John Gardner, *On Leadership*, New York: Free Press, 1989.

**Table B.1**
**Percentage of Officers Who Listed Various Leadership Qualities as Ones They Most Admire and Would Like to Emulate**

| Leadership Quality (with typical words and phrases in officers' responses) | Majors/ Senior Captains | Junior Captains |
|---|---|---|
| Character and Integrity: *honest, trusted, straightforward, genuine, fair* | 35 | 32 |
| Command Presence: *leads from the front, decisive, leads by example, charismatic* | 42 | 43 |
| Communication Skills: *listens, articulate, gives clear and concise guidance* | 34 | 12 |
| Composure: *calm, level-headed, cool at all times* | 7 | 15 |
| Develops Subordinates: *mentors, motivates, empowers, spends time with troops, delegates* | 27 | 26 |
| Empathy: *caring, patient, respectful of others, understands others' needs* | 38 | 27 |
| Intelligence and Mental Agility: *thinks on all levels, intelligent, resourceful, creative* | 26 | 30 |
| Physically Fit: *dynamic, energetic, fit* | 8 | 11 |
| Skill in Dealing with People: *approachable, personable, friendly, fun, candid, humble* | 31 | 33 |
| Task Competence: *tactically and technically proficient, knew doctrine, uses common sense* | 39 | 33 |
| Warrior Ethos/Professionalism: *courageous, mentally strong, determined, sets high standard* | 41 | 52 |

in each quality. This is obviously not an exact science, and others may use different groupings or may categorize specific words or phrases differently. The key point is that today's junior leaders appreciate far more than just tactical and warfighting competencies in their leaders. Not only do they recognize the importance of empathy, caring, patience, and listening and speaking skills, they often say these are the qualities in a leader that they most admire and that they wish to emulate.

After listing the qualities in a particular person that they most admired, officers were asked what position that person held when they knew or served with him or her. The results are listed in Table B.2. For both groups of officers, they were most likely to say that person was one echelon up. The most common response for the majors and senior captains was battalion commander, and the most common for the junior captains (they were platoon leaders and company XOs in their previous assignment) was company commander. Many respondents listed more than one position, and in some cases it was clear that they were describing two distinct leaders (for example, "battalion commander and platoon sergeant"). In other cases it was less clear (for example, "battalion XO/S-3").

**Table B.2**
**Percentage of Respondents Whose Most-Admired Person Held Given Position**

| Position | Majors/ Senior Captains | Junior Captains |
|---|---|---|
| Brigade Commander | 16 | 1 |
| Battalion Commander | 43 | 26 |
| Company Commander | 7 | 35 |
| Staff Officer (Brigade or below) | 20 | 23 |
| NCO | 6 | 9 |

APPENDIX C

# Lessons Learned by Junior Officers from Good and Bad Examples of Army Leadership

To focus the officers' thinking on issues broader than tactics and training, we included three items at the beginning of the questionnaire that asked respondents to 1) list the characteristics of a leader they admired; 2) state what position that person held; and 3) describe lessons learned from good or bad examples of leadership that they have experienced in the Army. This appendix provides the written answers to the third question, organized by major themes. To maintain the anonymity of both the respondent and the unit or leader being described, we made minor deletions in a few responses. Otherwise, the responses are presented essentially verbatim, with only minor punctuation changes made for clarity.

It is important to note some cautions regarding interpretation of these responses:

- First, these comments should not be construed as an assessment of Army leadership. The comments are representative of what the officers who completed the survey think is important in the way of leader qualities, but the comments are not necessarily representative of the leaders those officers had. For example, an officer might have had several excellent leaders but learned an especially salient leadership lesson from a single poor one, and that is the lesson that is reported.
- Second, negative comments (e.g., "Don't micromanage") do not necessarily indicate that an officer observed the bad example. He

or she may have had very good leaders and simply expressed the lesson learned as an admonition. The opposite is also true: An officer might have had or observed a poor leader, but expressed the lesson learned as an affirmation (e.g., "Trust your subordinates").

- Third, the officers may not necessarily have experienced something firsthand; they may have observed the good or bad behavior toward others. For example, an officer might have observed how a platoon sergeant acted toward his or her soldiers.

In short, we caution against use of these responses to prove anything regarding the nature and character of the Army's leaders. The value of these responses lies not in any quantitative interpretation, but rather in their demonstration that junior and mid-grade officers are learning how to lead from their own leaders. These responses also indicate what junior officers generally believe to be the most important leadership lessons. Officers preparing for command can enhance their self-development by reading these responses from hundreds of Army captains and majors, many with significant combat experience. They can use the responses to reflect on their own experiences, command philosophy, and approach to leading troops.

To aid in that self-development process, we have identified a few key themes that emerge from the majors' and captains' comments. The most general theme was also the most frequently listed—leading from the front or leading by example. This could mean many things to many people. To some it may have a literal meaning. One officer wrote, "I believe if you take the literal sense of 'lead from the front' many people will end up dead." A few officers mentioned being physically present or enduring the same physical hardships as subordinates. In the broadest sense, it shows that leaders are under constant observation by their subordinates and that they are teachers by example—for better or worse—whether they embrace that notion or not. Forty-four officers (11 percent of the usable responses to this question) used the phrases "lead by example" or "lead from the front" or a closely related phrase to describe a leadership lesson they had learned. Others did not use those phrases but expressed the same idea: "I have been able to watch what not to do and what influences others negatively" and "Soldiers are always watching and want to emulate their leaders." Twenty-six officers (7 percent) used some variation of the phrases "Do as I do," or "Don't say one thing and do another" as a leadership lesson.

A related theme is character. Twenty-five officers (6 percent) used the terms "character," "integrity," "honesty," "ethics," "do what is right," or some close variation. Five others used the words "moral" or "morals," which convey a slightly different meaning from integrity or ethics, but in most people's minds are probably very similar.

Forty-five of the captains and majors (12 percent) specifically used the words "care" or "caring" to describe a leadership lesson they had learned. Fourteen more (4 percent) used the word "respect" in the sense of showing respect for subordinates. Many more responses did not specifically use the words "care" or "respect" but still captured this sentiment, as in, "I learned to always put soldiers first, show genuine concern for them as individuals and they will run through walls for you." A closely related theme is not embarrassing subordinates. Thirteen officers wrote that leaders should not berate, belittle, or humiliate subordinates, and eight of the thirteen added that this should not be done in the presence of others.

Another common theme is trusting junior officers and giving them the leeway to make decisions and do their jobs. Thirty-seven of the captains and majors (9 percent) specifically used the words "micromanage" or "micromanagement" as admonitions against poor leadership. Nine more used the word "trust" to describe a commander's attitude toward his or her subordinates, as in, "Trust your subordinates, allow them to be the leaders that the Army needs." Other responses do not use the terms "micromanage" or "trust" but have the same sentiment, for example, "Give subordinates task and purpose and right and left limits, and they will perform and grow."

Two meta-themes emerge, both consonant with leadership theory more generally. The first is that leadership is inherently a social process requiring empathy, ethical treatment of others, and people skills. The second is that leader development is likewise a social process, since the junior officers are learning leadership lessons through observation and interaction with other leaders in their units.

In the following quotes, we present a handful of the written answers, representing each major theme, followed by the rest of the answers. Some of the answers relate to more than one theme; many do not relate to one of the themes we have highlighted but are still useful and insightful; a few, frankly, are not.

## Theme: Leading by Example, Leading from the Front, "Do as I Do"

> Don't "do as I say not as I do," show up on time, and don't distance yourself from subordinates.
>
> Bad leaders make subordinates do things that they were unwilling or unable to do themselves. Good leaders show incredible courage in the face of extreme danger.
>
> Say what you'll do and do what you say. Lead by example. Set high standards and demand people meet them.
>
> The importance of presentation, how you conduct yourself, is important in front of your subordinates; it is directly linked to their confidence and willingness to follow.
>
> Good leadership does not get any type of extra privilege when down in Iraq. For example, if the soldiers don't have air conditioning in the building they are occupying, neither should the commander.
>
> Good leaders create the desire for hard work in a unit. It is more than motivating others, it is the personal example that makes others want their leader and the unit to succeed.
>
> Leadership through intimidation works temporarily. Leadership through setting the example and team-building has lasting effects.

You lose your soldiers' respect the moment you are not willing to hold yourself to the same standards. Soldiers don't forget or forgive cowardice and incompetence.

I've seen officers who have used their status to get special privileges. I emulate the opposite, I don't ask subordinates to do anything I haven't or wouldn't do myself.

Soldiers are always watching and want to emulate their leaders. One of my commanders would say one thing and do another. This lowered his credibility.

## Theme: Character, Integrity, Honesty, Morality

Always be up front and honest with your soldiers.

If a person has no moral compass or principles, no integrity or character, they simply cannot lead soldiers.

Be decisive, honest, and do what is right. Discipline is the key to a good unit.

Do not try to change the attitude of a unit overnight, stand up for what you know is right all the time. If you make a decision you know is right don't waiver.

Always does the hard right instead of the easy wrong. Even if no one is watching because it will get back to your boss.

Good examples were those who always accepted responsibility and served others. Bad examples were those who passed the blame, had a low level of integrity, and talked about others behind their back.

Best to be open and honest with policies and actions.

Loyalty goes both ways, from subordinate to superior and back. Never compromise your beliefs or values.

## Theme: Care and Respect for Subordinates

There is a time and place for yelling/berating; if you do it all the time people write you off. Treat everyone in a way that makes them feel that they have input and can be a part of the solution.

I've learned to never belittle or chew out another fellow officer in front of his or your subordinates. When it comes to completing the mission you need to leave personal differences between others at home.

Lessons learned from bad leadership is not to embarrass subordinates in front of others. It just breeds distrust and drives down the motivation of the individual.

Do not ride your subordinates into the ground or belittle them. Give them the proper guidance and motivation then step back and counsel as necessary.

I learned that if you don't respect those around you, you will not have the support you need. If you don't ask for what your subordinates need no one else will, you take care of them and they take care of you.

My last Battalion Commander inspired loyalty in all of us. He took care of the soldiers and defended them. He inspired his soldiers like a good college football coach. He was at all training and made the whole Battalion feel like one solid team.

When you give punishment or reward, do so honestly, evenly, and with care. Take care of soldiers and their families' needs, the family is what keeps soldiers in the Army.

Must really care for people to truly lead them effectively.

Never talk down to others. Never attempt to make others feel as if they cannot add value because their experiences differ from yours (i.e. Airborne, Air Assault, EIR, CIB, etc.).

Good: The ability to relate with soldiers' issues, initiative, public encouragement, private personal scrutiny. Bad: uncontrolled emotions, poor tact, embarrassment of junior officers.

Poor examples are leaders demeaning others. Good leaders try to understand situations, rather than jumping to conclusions. Good leaders remember names and family information—birthdays, etc., kids' names. It shows that they care about the soldier as an individual and that promotes a more productive person.

## Theme: Trusting Junior Officers, Giving Them Leeway to Do Their Jobs, Not Micromanaging

Good: Trust subordinates to accomplish the mission with clear, concise guidance, intent, and end state. Bad: constant micromanagement.

Bad leaders tend to micromanage subordinates; by not de-centralizing the decision-making responsibility they develop weak leaders who are unable or unwilling to take charge. Incompetence also seems to be another aspect of a micromanager's leadership style.

One lesson in particular I've learned is that it pays to follow up. Not micromanage, just follow up to ensure accuracy and efficiency.

Good—Trust your subordinates, allow them to be the leaders that the Army needs. Bad—Crushing initiative, being abusive, or self-serving.

Every bad leader I have ever worked with never knew how to empower his or her subordinates. I have learned that leaders require the aid of the entire unit in order to be successful.

You have to give subordinates room to learn and make mistakes in combat. Mitigate the risk through continuous training and AAR.

Good leaders seem to empower subordinates in a manner that allows them to get their point or mission accomplished via the subordinates' initiative and not a lot of directives.

Good: prove a subordinate is worthwhile by dedicating time to him to explain a duty, express what a final product should include, and allow for personal initiative to complete it. Bad: no direction, unhappy with products, little feedback.

1. Officers must step up and make decisions. You are the coordinator for your subordinates and it is your job to ensure they know what you want accomplished. 2. Do not limit your soldiers' creativity and initiative. Tell them what you want done and let them figure out how to do it.

Micromanagement leads to poor performance. When you are operating 24-7 covering a large amount of ground. Sometimes you have to accept that a subordinate might do something a different way than you would. Tell soldiers what and why to do something and let them come up with how.

## Other Responses

Bad examples include directing subordinates to do something you would not do yourself, not confident, no sense of humor, and not doing your job.

"Do as I say, not as I do," abuse of power, and not knowing your job fully are all bad examples.

Good—do the right thing. Bad—higher command doing something they have told lower command what they can and cannot do.

Don't be a hypocrite, hold yourself to the same standard as your soldiers.

Do as I say not as I do (bad example/poor leadership) is a bad example. Leading by example is a good one.

I learned not to micromanage, "trust but verify." Good leaders lead by example and empower subordinates.

A bad example is doing things opposite of what you preach. A good example is making good decisions, taking care of soldiers for their own sake and leading from the front.

From Bad: respect is essential, humiliation in front of soldiers not a good way to develop junior leaders; if you want to challenge/develop subordinates by giving tasks without specific guidance so he can develop initiative/be creative, don't be angry if methods/results are different and criticize. Use exact guidance if mission is essential or mentor through process otherwise.

Lead by example, be the standard-setter and give your subordinates something to see; that is where integrity comes in. Always do the right thing no matter who is watching.

Impatience: Can cause a unit to lose morale and its overall focus. Trust: if you cannot trust your subordinates, their mission accomplishment is not going to occur. Loyalty: loyalty to your unit can mean the difference between life or death. Soldiers will give loyalty when shown it in return.

Good: don't be afraid to delegate and allow subordinates to execute off of intent. Bad: uninvolved, afraid of confrontation, never made a quick decision. Good/Bad: don't be afraid to "be" a soldier while leading them, just don't forget to lead. Also prone to outburst over info never relayed in the first place.

A bad leader, one who micromanages to the fullest, and blows up at the slightest inconvenience to himself, is one who you can anti-emulate. However, I learned that if you give people your trust at least in the case of my awesome CO, those people will desire strongly to perform to their best ability, every task laid before them. I know also that while tactics, competence, etc., are key ingredients, often times interpersonal skills are absolutely essential. He also taught me not to let people take advantage of me or let them "walk" on me, but rather, to be much more assertive. He came to my wedding even, two months ago.

Good: even a COL can learn something from a private, listen to what everyone has to say; set clear boundaries for subordinates and let them execute; and treat every subordinate as a resource you are trying to get the most out of. Bad: leaders should never talk negatively about, well, almost anything, in front of their subordinates; leader needs to know his job or he's wasting everyone's time; wasting soldiers' time you either work, train, or go home, hurry up and wait should only happen in the field.

Good leaders always lead from the front, treat soldiers with dignity and respect, and ensure that every measure is taken to keep them safe. Bad leaders have a "me first" style, are egocentric, and have poor management skills.

Don't humiliate people in public, don't waste people's time, be honest, fair, and tell people how it is going to be.

Don't expect of others what you do not put forth yourself. Calm leaders create calm subordinates (overreaction under pressure from demeanor is not needed). Listening to others is a good indicator that you care.

Bad leadership is talking down to/about your subordinates, micromanaging. Why would they want to perform for you? They are not going to listen to you and they will criticize your judgment.

Do not "shotgun blast" everyone for one person's mistakes. Do not micromanage subordinates. And do not manifest, by words or deeds, distrust in those that work for/with you.

Be willing to do (and capable of doing) anything you ask your men to do. Do not micromanage.

You have to care about all of your soldiers, regardless of rank. As an [position] I was diagnosed with [disease], everyone except my commander asked me how I was. He preached constantly about taking care of troopers but didn't do it himself.

Give subordinates responsibility. Don't look over their shoulders all of the time. Do what your troops do, don't make them do all the work. Listen. Be a physical example.

Good leaders set and follow examples, stand up and take care of soldiers, and take charge. Bad leaders spotlight leadership, are promoted not based on merit, just checks the box.

Treat soldiers with respect always. Set the example in all you say and do.

Generally I've learned that being yourself, always leading from the front and always treating people with respect is most important. If you don't or can't do that you are probably a bad leader.

Don't belittle subordinates in front of others, adopt a "Do as I do" kind of mentality—lead from the front. Some people do discriminate against others with different personality styles (i.e., extroverts vs. introverts).

Trust, until proven untrustworthy; have integrity, always do your best not just enough to get by; and have respect, initial stereotypes quickly dissolve once you work with one another.

Lead by example, be ethically centered, don't have a "check the block" mentality. Don't focus decisions on personal career advancement, take calculated risks.

Bad lessons: favoritism, poor fitness/composure, stresses subordinates needlessly through over-work and micromanagement. Good lessons: lead by example, compassion for soldiers, maintaining calm composure and emotional maturity.

There are good leaders and bad leaders. There is not much in between. Good: lead by example, give the freedom to make mistakes (Division level). Bad: Micromanagement, no senior rater development, no freedom to make mistakes (Battalion level).

Set the example, lead from the front. Communicate to subordinates, trust them, and provide top cover.

Must have a presence with subordinates, be open to conversation and criticism, must do what your subordinates do as an example.

As a leader you must always lead from the front, don't ask a soldier to do something you wouldn't do yourself.

Take care of your personnel, bestow trust, mentor, coach, teach. Someone is always watching, don't compromise your integrity.

I have learned from several bad examples not to micromanage and never order someone to do something you are unwilling to do yourself. From good examples I have learned to treat soldiers with respect and always be present as a commander.

I learned to set example in character traits and to demonstrate by your actions that you care about their/your own family.

Bad examples: refusal to listen to subordinates/take their criticism because they are subordinates. Refusal to explain the "why." Demanding mission completion without giving the tools. Micromanaging. Good: a soldier that knows "why" can make things happen. State your expected standards up front then correct the behavior. Everyone has a breaking point—watch for it. Soldiers who feel respect return respect.

Care for soldiers and be honest.

Bad—A leader that does not have trust in subordinates to have success (micromanaged is a morale killer). Good—A leader that empowers NCOs and small-unit leaders to do their job with guidance and interest. Shows up and participates in training. Is concerned about his men and will stand up and protect his men if they need him to.

Good leadership: look out for your soldiers, help mentor your soldiers, listen to your soldiers, lead by example, and train to standard. Bad leaders: very little interaction with troops, little knowledge of doctrine or tactics, doesn't push him/herself, doesn't lead by example, and treats subordinates with little respect as though he/she is better than anyone else.

As a leader, traits I want to integrate into my leadership style and example of what not to do when in command.

Having a commander with a knowledge of all systems in his company. Leaders setting examples in all aspects of training. Not taking advantage of rank held over subordinates, and willingness to not take comforts the soldiers don't have.

Never order soldiers to do something you are not prepared to do yourself. Actually, intently listen to people like they are the most important. Attack problems from all angles if you have to in order to get to objective. Stay calm. Good training is taking care of soldiers, not giving them time off. Do not let risk management inhibit tough, realistic training. Train as you would fight in combat; don't just give it lip service. Use time efficiently through good planning and preparation.

People that always have to have the last word generally are not good leaders. The best leaders are fit and lead by example.

Lead by example, always give 100%, don't complain to subordinates, be technically and tactically proficient. Train to standard through hard, realistic training scenarios.

Good: Be present at training, lead by example, develop junior leaders, don't be critical. Bad: Sharp-shoot peers, physically unfit, too much time in the office and not with unit.

Good: Lead from the front, also be the first out the door on all Airborne operations. Bad: Profile could not wear IBA [individual ballistic armor] but still was in leadership position in combat.

Always lead by example and from the front.

Lead from the front on all things, civilian and military. Do what is legally, ethically, and morally correct. Set the example, maintain self-control at all times.

I learned that good leaders must endure the same hardships as soldiers to be respected, must work harder. Good leaders separate self from subordinates off duty, always set best example, and must prove your worth.

Leading by example and giving subordinates the freedom to learn and grow professionally.

Bad—Don't trust subordinates, don't listen to others, is always right, their way and none other, bullies and rule by intimidation and no toleration for opinions, dissent, or advice. Good—See answer to question #1 [in response to question #1 asking for characteristics of a leader the officer admires, the officer listed competence, taking care of soldiers, empowering soldiers, mentors, accepts but fixes mistakes].

Good: Wants to be part of the mission. Cares for soldiers' welfare. Predictable and easy to communicate with. Bad: Never goes out on patrols unless influenced by higher. Doesn't fully understand mission because he doesn't get out enough. Doesn't set the example ("Do what I say not what I do").

Good—set high but realistic standards. Work hard, not long when possible. Set the example. Bad—know your job. Screaming doesn't work. Be approachable.

Good examples: I learned the value of knowing/reading enforcing doctrine. Bad examples: not leading by example.

Be decisive, don't denigrate subordinates, be succinct in issuing instructions, and lead by example.

I've had or worked with leaders who only worried about themselves. Lead from the front and actions speak louder than words.

Be—have to be in charge when tasked. Know—have to attain knowledge to assist subordinates. Do—have a set the example for others to follow.

Don't be self-serving. Set the right example in all of your actions. Mentor your subordinates. Be patient.

You always learn from good and bad leadership and you always see both. I have always just made mental notes on how I would do things or how not to do things based on what I have seen.

I have learned that soldiers' perceptions of what you do are almost as important as what you do.

Good leaders know what is going on in their organization and interact with their subordinates. Good leaders show the subordinate what right looks like. They issue clear and thorough guidance.

I have been able to watch what not to do and what influences others negatively.

Often field-grade officers micromanage, they adopt a zero defects policy and overall can't relate to Company level officers. Bottom line . . . no mentorship.

Nervous/hesitant leaders sow the seed of doubt, quiet/shy leaders are not effective, overbearing/insulting leaders create animosity, and micromanagers create an environment of ineffective subordinates.

Engagedness—Bad lessons learned. I will not give my subordinates so much freedom as to give "no guidance" but not go the reverse direction and micromanage.

Good: How to present myself as an officer. Bad: What not to do as a commander in combat.

The good leaders show up to PT.

Set high standards, be a practical thinker, challenge yourself and exhibit your best efforts always, and learn from mistakes.

Be seen by the soldiers in your command (especially in combat). Making a decision in combat (even if it is the wrong decision) is better than inaction. Be firm, fair, and honest with your soldiers. They will smell a fake a mile away.

If you are in a leadership position then act like it. It does not matter what type of responsibility you have.

I have learned the most lessons from bad leaders. I see them and their actions and it reminds me of how poor some leaders are and it motivates me not to be like that.

Have to suffer hardships alongside your men/peers. A little preparation goes a long way.

Had a company commander who only cared about getting a top-block OER. He taught me whole lot about what kind of leader not to be.

You cannot fool soldiers by saying the right things but not putting things into practice. Positivity breeds positivity; negativity is a cancer. Discipline is the key to success at every level.

Observed what not to do from examples of a company commander that I worked for as an XO. Opposite of the above qualities [in response to question #1 asking for characteristics of a leader the officer admires, the officer listed leads by example, soldiers' needs before own, tactically and technically competent]. Even in combat this individual consistently provided poor examples. I have tried to emulate the extreme attention to detail of a previous battalion XO.

Don't make changes impulsively (ethical behavior changes aside). Include others in decision-making process, which leads to changes as much or as often as possible. Ensure you know your unit and the people in it. Never ask someone to do something that you wouldn't do yourself.

Bad examples: overly confident, poor social skills, derogatory towards subordinates/peers, fails to set the standard repeatedly, sets double standard, lacks vision, drive and purpose, and is selfish.

I learned to always put soldiers first, show genuine concern for them as individuals and they will run through walls for you. Always be willing to listen to advice from up and down the chain of command, then be open-minded enough to act on it.

The biggest lesson I have learned is to do what is right for the troops. This does not always mean what makes them comfortable and happy, nor does it only refer to their direct actions.

Always visit your soldiers. Remember soldiers with everything you do and plan. Remain calm, never make an important decision when frustrated or angry.

A good leader not only tells you what to do but teaches you along the way.

Treat subordinates as human beings and not as paid workers.

Praise in public—be critical behind closed doors. Don't be afraid to admit you don't know something.

Always be willing to learn and share experiences. Be patient with subordinates—they look to you to be calm during stressful times.

Be aware of your command climate and what is going on from PLT [platoon] to the level just below you (be seen). Always seek self-improvement in one area or another.

As a leader, you are always on display. Everything you do is looked at, analyzed, and criticized. The single biggest example I can think of is separating personal from professional. I once had a LTC say to me, "I take this as a personal affront." It was a mistake I made and deserving of punishment, yet not to the level that it was taken. So, as a cadet, I learned very quickly to separate my personal feelings from my professional decisions in that regard.

Actions speak—must do what you say.

I have learned that there is a distinct difference between being a leader and being an effective leader. Effective leaders are personable, goal-oriented, detail managers who find creative ways to motivate their subordinates and empower their unit to actions. Effective leaders are dynamic, charismatic, moral, and loyal.

Leadership goes beyond what happens at the "office." Leaders are human and make honest mistakes but a leader's character must be consistent in personal and professional roles. The leaders I admire most held a consistent bearing in and out of uniform. This is especially important when times get tough and stress builds in a unit. Good lessons were always focused on the mission and the best interest of the unit. Bad lessons were focused on an individual goal or desire.

"Do not command what you cannot enforce." You must go and see things for yourself. Success is its own justification. You can do it all for a short time, endurance is built on the training of subordinates to do it themselves. Be yourself. There's only two reasons to do things—1) it's the right thing to do, 2) someone gave you a lawful order. It is your duty to correct people (superiors, peers, subordinates) when they are wrong.

Poor leadership wastes my time. Poor leadership does no work on their part—they push off tasks on subordinates with no real guidance, but then let you take the blame when standards are not met. I learned to CLEARLY tell my subordinates what we're doing, why we're doing it, and when we will finish. And I don't disappear while they work. I am available and coordinate for whatever they need.

Bad examples: treat person with a level of professional respect commensurate with their rank., don't talk to a CPT like a LT; assume he knows what he should know until he proves otherwise. Micromanagement is the tool of an ineffective and insecure leader, train your subordinates to accomplish their assigned tasks and give them the leeway to accomplish it.

Many, but maybe the most important is to be able to balance work and play. If you show the soldiers that you will work hard right along beside them they will respect you more. Also, showing them that you are human and like to have fun just as much as they do will also help strengthen the team.

Lead by example. Hold yourself to same standard as your men. Humility. Confidence. Those that possessed these qualities and demonstrated them were good examples. Those who did not have these qualities or did not demonstrate these characteristics were bad.

Good: Lead by example. Attempt to do what is right always, even off duty. Apply common sense to your actions and decisions. Bad: Keep soldiers at work until the leader is ready to go home (wasting soldiers' time). Make decisions based on what is good for you not for the unit or soldiers.

Listening—I had a commander who didn't really listen to peers or subordinates. He was determined to do what he wanted. Could not lead from the front. I thought it was just slang or a slogan, but if you can't perform subordinates will view you as incompetent.

Not supporting decisions made by subordinates. Superiors not counseling officers, but demanding them to counsel. Leaders using a generic bar for reward/punishment, not taking into account personal situations. Keep the family informed. Include the chain of command in decision-making, officers need to get NCO input. Officers doing stuff themselves, not delegating, while senior NCOs delegate with no mentoring = creating a weak lower enlisted rank. Doing stuff with the unit and soldiers—not making excuses to miss training.

Lead by example in everything you do. Give specific guidance to subordinates and show a genuine concern for the delegated mission. Be more concerned about the mission and welfare of troops than own needs. Soldiers know when you're not. Do all the things in question #1 [in response to question #1 asking for characteristics of a leader the officer admires, the officer listed competence, understanding people, empathy, communication skills, command presence, and leading by example].

Bad: Explodes (i.e., berate, belittle) subordinates when they make mistakes. Does not provide an atmosphere for learning. Takes too long to make a decision which affects his subordinates' livelihood or development.

"Easy-going" leaders don't always make good leaders. Put your soldiers' needs above your career.

Always make time for PT, work can wait. Actually have genuine care for your soldiers, they can tell when it is fake. Finally listen to ideas from everyone and incorporate them.

You must take care of soldiers. If you do then they will take care of you.

Take care of your soldiers, be positive about your decisions and stick to them and be prepared for anything. Take your job seriously, don't do half a job.

Do not put yourself first above your troops, soldiers quickly recognize leaders who care about themselves more than the unit. Time management is key, troops become very frustrated when timelines are not met and tasks always take longer to accomplish.

Patience is important (especially with COIN [counterinsurgency] environment of today). Be able to "walk the talk."

Not adhering to standards that you set but expect juniors to follow.

Good—places mission over personal desires. "Takes care of soldiers." Bad—says to do one thing, does another. Personal goals put in front of mission accomplishment. Not willing to make personal sacrifices.

Bad examples: poor communicator, couldn't describe intent to subordinates; placed his career and OER above welfare of troops; dictated standards that he didn't follow; and questionable moral standards.

Bad examples: Quick to judge, doesn't do what soldiers do at PT and training, afraid to make decisions in combat because of risk to soldiers, and blames everyone else but doesn't take responsibility when things go wrong.

Good—all people are human and with that we all make mistakes. However as leaders you have to raise the bar a little higher than your soldiers and learn from the mistake and not make it again. Bad—not keeping your word and not leading by example. (i.e., just don't say it, do it).

Lead by example. Be at training—all training. Do not lie, fabricate results, fail a PT test, or be overweight. Every soldier watches your actions and interprets them to suit their own agenda. Talk to the soldiers/leaders regularly—formal and informal—take their input to discuss it with your leadership and give feedback to the unit. Nothing makes the soldiers happier than to see their ideas put into action. Leading is never about you—convey that in words and actions. It's all right to make mistakes—admit it and move on. Same applies for your soldiers/leaders. Integrity is non-negotiable. If you make a controversial decision, make it in the benefit of your soldiers. Most often better to watch and listen than to speak up—stifles innovation. You must have 2-way communications with your boss—he/she needs to hear good and bad. A good leader also serves as a mentor. Give subordinates room to grow, learn, make mistakes. Be technically proficient—back up what you do with TMs, FMs, AR, ARTEP, etc.

Not to waste soldiers' time, if they don't need to be there then let them leave, especially with today's OPTEMPO. Always push your soldiers hard during training.

Communication between a company commander and 1st sergeant as well as between battalion commander and CSM [command sergeant major] are the linchpins of an effective unit.

You can be a half empty or a half full kind of leader. Subordinates of half empty will always question their judgment and learn to not take risks.

Bad lessons: unrealistic standards, making decisions personal not professional, did not take other options into account (my way or the highway).

Leaders who take the time to get to know their subordinates personally are more effective. Those who rely solely on their rank to maintain control are less effective.

I have learned to apply common sense to all situations. If an idea does not pass the common sense test, you must remember that a soldier is executing the associated tasks—it is not just a slide in a show or a block on a calendar.

You have to take time to talk with your subordinates regularly if only for a few minutes a day.

I learned that leaders are highly dependent on their own experience and knowledge. However, the more they lack these, the more aggressive their personality must be.

Bad example: Battalion Commander seemed to have no idea of what was going on and no one would follow her. Good example:

Battalion Commander just knew about what was going on in his organization.

Never make decisions without consulting others if time permits. There is no such thing that "can't be done"; there is always a way. Loyalty is a two-way street, both up and down the chain of command. Teamwork is key at all levels.

Bad leaders always rely on NCOs (I knew Platoon Leaders who didn't rely on their NCOs' experience), won't talk with you about anything. Good leaders are willing to talk.

Always keep everyone informed, cross-talk problems to get solutions, and make a decision—do not let anything fester or build.

If you give a subordinate a task, monitor them but allow them to do the task without doing it for them. When in charge don't forget how important everyone else is to the whole.

Do not assume your subordinate leaders are incompetent. They will lose trust in you.

Give subordinates task and purpose and right and left limits, and they will perform and grow.

Learned how to/not deal with subordinates. Count on those under you to get the job done. It's their job. Never let someone slide, the standard should always be maintained.

Good: Battalion Commander kept good situational awareness while letting his company fight and developed the situation, gathered information and made decisions.

How to communicate verbally better, be careful what you ask for, and be aware of what others think of you.

I have seen the worst company in my battalion change into the best, overnight, solely because of a new commander. The actions of a unit are directly linked to the leader.

It's better to make a decision, whether right or wrong, than to not make one at all.

I've learned not to be petty and avoid knee-jerk responses. I've also learned that the most important thing isn't always the mission. Your soldiers are the most important part about your job, always.

I learned to always take care of soldiers, know your job, and motivate soldiers.

Good leaders take care of their troops, bad leaders take care of themselves.

Be personable with your subordinates. No matter what they may say about not wanting a visit, show them that you care and are grateful for their service.

Always take care of your subordinates; manage your own career; don't expect others to help you; be confident and self-sufficient.

Soldiers can spot a fake from a mile away. Care about your soldier, keep yourself physically fit, and develop your soldiers as warrior leaders.

Understand people, respect them, and soldiers respect you. Train hard, give soldiers a chance to lead and follow. Mistakes happen, how you recover from them is what sets you apart.

A bad example of leadership is someone who could not identify or understand what subordinates needed. They put their careers and well-being ahead of subordinates.

You can't fly off the handle every time you get bad news. Eventually people stop telling you things that you need to know. You need to engage soldiers. Don't rely on what you hear, get out there and show them that you care and listen when soldiers talk.

An example of bad leadership is micromanagement, assigning same tasks to multiple officers.

One company commander was so worried about looking bad to his higher that he micromanaged my platoon to the point everyone hated him. One company commander seemed aloof about what was going on to the point where soldiers asked me, the XO, what we should do.

An example of a bad leader is one who adopts the "what have you done for me lately" attitude, severely micromanages.

Allow your subordinates to develop, don't micromanage them. There are usually multiple ways to solve a problem. Learn how to motivate your soldiers, everyone is different and may need to be approached in a different way.

A commander can tear down a company if they do not have trust in the leaders of the company. This also cripples leader's development due to the fact that the commander stays in your business.

Good leaders never underestimate the value of listening, they trust their subordinates to execute (avoid micromanaging) and compliment people when they do well. You should always maintain control of your emotions, avoid hypocrisy, and be flexible.

Allow honest first-time mistakes. Allow soldiers (officers) the ability to accomplish the mission without micromanagement.

Positive results never come from micromanagement or lack of trust.

I learned to trust your instincts and be decisive, to constantly seek more responsibility and ask questions, and to care about your peers and subordinates.

Don't be afraid to say what you think, be with your soldiers whenever possible, think out solutions to problems, don't have knee-jerk reactions, mentor and counsel your subordinates. Take care of soldiers and their families.

I learned not to do bad things—things that had been done to me—people not caring about me just wanted me to do the job and that's it. I learned to depend on people as a team, and that everyone's input is important.

Bad Example: Not caring enough about the soldiers' welfare or living up to Army values. Good Example: Leaders going beyond the bounds of duty to help out soldiers.

Bad: Not treating soldiers with respect (as a person), letting young 2LT [second lieutenant] be more of a hindrance than an opportunity to teach. Good: When figuring out development, taking into account as much as you can about soldiers' family situations.

Good—develop relationships, truly know your soldiers. Have fun in what you do. Bad—don't belittle or demean when disciplining, don't blame others for your mistakes.

Be on time—right place at right time in the right uniform. Take care of family. Be technically and tactically proficient. Have situational awareness.

How to care for soldiers and the importance of their families as part of a team. Being proactive but not falling out of the foxhole in anticipation of an event.

A caring approach to soldiers and their families is important. Strict adherence to standards is equally important. Hot and cold leaders that are unpredictable are bad in peacetime and in war.

Know your subordinates, be technically and tactically proficient, and be flexible. Pay attention to duty, honor, and country. Care for but don't baby-sit your soldiers, be prepared or ask questions if you do not know something.

Care for the individual soldier and things will work. You make mistakes, learn from them. Tell people all that you can, if not they won't commit.

Treat people with respect, just as you would like to be treated. This works far better than creating an atmosphere of fear and tentativeness.

Respect the soldiers' time always, do not waste this most valuable of resources. Always instill a sense of purpose in your people for everything you all do. Be well rounded, the Army isn't everything.

The lack of courage or willingness to support soldiers, NCOs, and subordinate officers will destroy both loyalty to the chain of command and morale within a unit.

Good leadership grows from selfless service. Anyone who is purely self-motivated will ultimately not be effective. One must have a positive attitude—a can-do attitude and the ability to focus without allowing emotion to control one's actions.

Deal with issues in a timely manner. A leader may be the one who is responsible but it takes everyone to be successful. Good leaders take the blame for failures, but attribute successes to everyone else.

A private will stand in shoulder high freezing water if you tell him to—but he will appreciate you and the situation better if you tell him why he needs to stand there.

Good Examples: time management, delegation, presence on the battlefield, caring attitude, concern for subordinates, clear understanding of subordinates' workload. Bad Examples: unwillingness to listen, not communicating purposes to subordinates, being unapproachable, arrogance based on rank, unreal outlook/perspective.

Good—Key leaders taking time to speak with subordinates about important issues, promotion, family, etc. Bad—Leaders are not entitled to BAD days, e.g., my Brigade Commander chewing me out because he was mad at someone else.

Good leaders seem to be able to connect with their soldiers in a manner that commands their respect. The best leaders are knowledgeable in their field and can apply that knowledge. They often seem unfazed by events around them and never lose their bearing. They also seem to care for their soldiers. Bad examples of leadership I have seen in the Army include leaders that cannot control their emotions, who complain. Who lack empathy for their soldiers, and are stuck to a rigid doctrine during operations.

Good: take care of soldiers—they will take care of you; strict accountability of self, soldiers, equipment; to be fair and open-minded. Bad: poor communication—withholding important mission details, one-sided discussions, and always group punishment.

I believe if you take the literal sense of "lead from the front" many people will end up dead.

Be explicit and direct when conveying your intent, be willing to listen, everything is a situation, always get both sides to the story.

Self-serving leaders are easy to spot. They don't help move things forward. Knowing your organization with a high level of detail is often hard, but it is extremely important.

There are different ways to reach soldiers. Take the time to find the way, the RIGHT way.

Always take a step back, think about a situation and then act. Be aggressive but understand the consequences of your actions.

Learned how to take care of junior leaders. Did not receive coaching or mentorship from company commander. Learned that it's always important to listen to all aspects of a situation before making decisions that can affect the lives of my soldiers. CO did not do this.

Listen to subordinates even if you disagree with them. Arrogance is not an effective form of leadership.

Be open, ensure your standards are understood, and hold soldiers accountable for their actions and what you tell them to do.

Backwards plan. Involve subordinate leaders in planning process. When something must be done, the battalion often cannot wait for the brigade to put out the final guidance, they must act.

Making the subordinate come first and truly mean it vs. lip service. Zero defects mentality. Micromanaging and taking on one of your projects and saying they can do it better. No end state. Listening to those in chain and other soldiers who may approach you.

From the bad = what not to do in a similar situation. How not to overreact and provide "top cover" for honest mistakes. Provide purpose, motivation, and direction and empower junior leaders. Understanding my way is not the only way. Accepting of 90% perfect when done by subordinates instead of micromanaging or doing it myself.

Assess unit before making changes. Never talk about your OER or need to make rank in front of soldiers. Never play favorites. Equal standards for everyone in unit. Never talk down to soldiers. Always show trust in subordinates and allow them to make mistakes.

Basically to be a good leader you have to show your soldiers that you are for them and you have to share experiences with them.

Good: The ability to see the "heart" of a problem without getting caught in the "weeds." The ability to influence people to give their best without threats or promised. Bad: Disconnected from soldier issues, opinionated to the point of intolerance.

As a leader you can't make everyone happy. You have to make the tough decisions which could affect the lives of your soldiers. At times you have to stand up and let your seniors know that they sometimes make bad decisions.

No one person has all the answers. A good leader knows how to combine all appropriate ideas and experiences to set up his unit for success.

Open and effective communication is key. Must clearly convey intent and assume nothing when it comes to getting your point across. Must be concise or else audience's mind will wander.

My last commanding officer. His way was the only way of handling a mission. He talked down to lieutenants.

Leaders who lack a full understanding of what leadership is and leaders who put their own career over the welfare of their soldiers are examples of bad leadership.

If you are more interested in yourself than your soldiers, it will be noticed.

Listen to subordinates' ideas, support their input. Use resources creatively, seek resources from non-traditional sources. Let people know how their effort fits into the big picture.

Be a listener, inclusive with shared personal concerns for soldier and family well-being, understanding but decisive.

Focus on the objective but do not sacrifice the living of subordinates en route to it. Listen and take recommendations but remember that the leader must be the decision-maker.

Bad: Constantly threatening subordinates, don't do it. Don't constantly change things for change's sake. Good: Enforce and strive for tough standards. Show genuine concern.

Yelling at subordinates who make "non-life-threatening" mistakes or publicly degrading a subordinate alienates the individual and often sways the morale of the entire unit.

You must never jump to conclusions and have as many facts as possible before making a decision. Never make your peers or subordinates feel inferior but a part of the team and let them know that their opinion counts.

Leading by intimidation and tirades is not effective.

Don't treat subordinates like crud; leaders must understand that this is a people-centric organization.

Bad—you can't make good decisions when you are emotional, demeaning others shows how shallow you are. Good—make sure your comments add value or don't add them, be consistent.

Bad—Self-centered people walk all over their people to achieve their own personal agenda. Leaders who cannot accept bad news. Leaders who are not willing to make a decision, not willing to lead.

Bad example: Company Commander who was short and verbally abusive to subordinates combined with tactical incompetence equaled no good leadership qualities.

You can't make everyone happy all the time, and be firm but fair with subordinates.

A bad example is someone who does not know the job and is overly reliant on others, acts like they know things they do not. A good lesson is to know what you are doing and ask questions to constantly learn and improve.

Bad: Never miss an opportunity to look good especially at the expense of or with the work of someone else. Good: Be the quiet professional who takes pride in a job well done, train and work hard, be generous with time.

Bad examples include those who look down on soldiers because of a rank lower than their own, and those with no interpersonal skills.

Good: Flexibility is key, knowledge of systems and tactics has no substitute. Bad: If you don't plan, you will meet with disaster every time. If you can't communicate effectively, you will waste everyone's time and look like a tyrant.

Don't fuss about things you can't change. Support higher's intent as your own, don't be political, don't let your buddy/peer fail.

Bad leaders have a temper, are never wrong, and believe no one is as smart as they are.

Good leaders did not jump to conclusions, mentored junior officers, and gave timely guidance and feedback. Bad leaders let soldiers know they were better than them because of their rank, jumped to conclusions, and did not mentor officers in new duties.

Good leaders know the correct level of management needed (micro vs. macro) and show great personal courage.

Bad leadership is that which takes away initiative from subordinate leaders, micromanagement and not treating others as fellow professionals in a field.

Distrust of subordinates and micromanagement are always counterproductive.

Bad leaders are rash—rush to judgment. Don't see the big picture—only what affects them. Micromanagers that don't see the wealth of leadership and competence that exists in subordinate leaders and stifles their ability to contribute by dictating everything.

Good—giving subordinates latitude to make decisions and execute them. Bad—micromanagement.

Cannot micromanage subordinates, let junior commanders learn from mistakes.

Ensure you give your subordinates clear task and purpose. Trust your junior leaders to do their job. Do not waste soldiers' time.

Good: Power of positive leadership, validation of information, motivational in communication. Bad: Losing temper and attacking character of officer, micromanaging and stingy on trust.

Good—empower subordinates to accomplish, based on intent and vision, always optimistic. Bad—micromanager, not open to outside ideas, pessimistic.

Bad: Being weak, not getting away from the computer, or using excuses. Being immature, not taking advice or learning from mistakes.

Not listening to feedback from subordinates can lead to repeating of stupid mistakes. Not effectively communicating one's intent can lead to confusion among soldiers.

You can't be soft. I had a soft Company Commander and even when I wanted to work hard for him I didn't know how.

Good leaders always try and improve your knowledge. They are always fair and equal to subordinates. Bad leaders do not make subordinates "buddies."

I learned patience, either on the battlefield or in garrison, being able to keep your cool and make sound decision was best.

Bad leadership has no will or desire for the job, requires no accountability and takes no responsibility for things gone wrong.

Dealing with people on a day-to-day basis well usually precipitates sterling results from the interacted.

Good NCOs are ones who look out for you, they can make or break a unit.

I have learned to listen and find the "fact" before I react. I have learned to make a decision because soldiers are depending on you.

Leaders must be present and engaging with subordinates, be loyal to soldiers first and superiors second, and they should also be decisive and inquisitive.

Good example: how to value, empower, appreciate, and reward individuals of all ranks. Bad example: how not to treat peers and subordinates.

Know your strengths and weaknesses.

GOOD: Willingness to move into harm's way with troops. BAD: Exact opposite of previous; blaming others for Co/Btry mishaps and shortcomings.

Don't ignore your troops. Get to know them, but don't become overfamiliar.

You can do the task or you can gripe about the task and still do the task. Army is as difficult as you make it.

I have learned to explore both sides of a story before disciplining soldiers, at times there is more to an issue than what is reported. Think before you act, do not act when angry.

Never think I am better than someone because of the rank I wear. Always understand the human dimension.

Bad Example: The leader who is not approachable because of a lack of knowledge. One who is career-oriented not soldier-oriented.

Do not promise things you cannot deliver. Mentor two levels down, conduct exercises and missions with subordinates (esp. in combat) to assess and help them develop.

Bad: to allow your subordinates to know that all you care about is to make the next rank, it really affects morale.

Telling all of his platoon leaders that they didn't have the ability to speak or address a large audience is an example of bad leadership.

Don't be too rigid, self-sacrificial, pay attention to detail, introspective—what can I do better.

From bad leaders I learned always value the opinions and suggestions of subordinates. Often it is supervisors who cannot separate their rank from their goal of achieving their mission who are difficult to work with/for.

There is a need to hold mid-level leaders accountable for bad actions (captains and lieutenants)

Rehearse and train as it will be my last time before going to war, conduct battle/combat training during PT or regular training day.

Do not base your opinion of someone off other's recommendations. Get to know the soldiers before making judgments. Instill discipline in yourself first, also professional development begins with the individual.

Give honest feedback, know subordinates on a personal level, have high expectations, communicate intent well and ask for subordinate input to solve problems, don't yell unless necessary, and determine priorities first.

Don't waste soldiers' time, don't treat younger officers poorly just because you are in a leadership position, get to know your soldiers and what motivates them, and train soldiers to the best of your ability.

The MOST important element of leadership is developing personal relationships.

Plan thoroughly, plan for contingencies, do not accept "no" from supporting units. Be persistent!

A leader must be open to criticism and recommendations to decide on the best course of action.

I admire people (leaders) who are charismatic, have heart, are physically fit, and have the ability to use common sense (an art form that you don't see much anymore).

Always look at a person's heart, not what patch he/she is wearing.

Total central dictator-style leaders kill morale and productivity.

Bad—One supervisor did not prioritize and constantly followed up on all tasks simultaneously. Good—talking with soldiers about their families/lives in informal setting.

Losing control before enough facts are available to make a good decision or understand what occurred.

Don't be a task-master, be a leader.

You must act decisively, however one must develop the situation first. If not, often one risks failure.

Never befriend your peers that are in your unit when you are the commander, this leads to favoritism.

The Army places OER/NCOER above all, makes it a focal point for subordinates. I have learned how to relate/deal with people on many levels.

Bad: Poor guidance/lack of communication, a poorly informed soldier will have lower morale and consequently do a poorer job overall.

Lessons learned are about 50/50 good to bad. Leaders who don't know what they want and make subordinates "spin." Leaders who cannot handle bad news.

This would take a book to answer.

Interpersonal skills are essential, you have to candidly tell people what is expected and provide feedback on performance.

Good lesson: Informal counseling that truly was a two-way conversation. Bad lesson: Threaten with relief one week into command over temporarily losing NODs [night observation devices] on an FTX.

Always have a plan, it's easier to change a plan than to create from scratch at the last minute. Be straightforward with people and they will follow because of trust. "Arrow of responsibility points down, not up."

Keep your cool, always do your best, know the book, and be tactically and technically proficient.

Be yourself, anything else is wasted energy. Prove yourself by what your are doing now not by what you might get for rewards later.

Bad leaders are indecisive, are non-confrontational, are lazy and impersonal. They are myopic and focused on methods versus end state.

First report is always wrong, don't jump to conclusions. "You may not be in the right place when you are with your men but you are never in the wrong place when you are with them."

Always be at the decisive point of an operation to enable your ability to assess and control the situation. Thorough planning of every action will always make your unit better prepared.

Good: give a little, but clear, guidance and the men will work wonders.

Yelling is not effective if employed often. Leaders must listen to subordinates. Knowing where to be (what activities, position on the battlefield).

Take responsibility and be accountable for both good and bad. Encourage open communication both above and below your level. Listen to others.

Bad examples—no social/interactive skills. Failure to relate. No visible involvement.

Rank doesn't equal right. Motivation—both the carrot and stick methods work, but too many people only use the stick method.

Bad leaders don't keep their troops informed, blame their subordinates for failure.

Large ego = poor commanders. Candor is not always rewarded.

Both good and bad leaders may experience success in the Army. Bad leaders tend to break or discourage younger leaders. Most great leaders possess a natural charisma that overshadows all other leadership qualities.

How to do and not to do things.

Some commanders who ask for too much information to make a decision.

Need to inspect mission critical items personally, and conduct rehearsals even though you have done the mission before.

Good: Understanding full picture before making decisions, handles bad news as challenges. Bad: Snap decision-maker, shoots the messenger vs. focus on problem.

Learned how to timelessly serve each and every soldier of the unit to give the very best support to the troops.

Bad leadership: had a Battalion Commander that was not tactically competent. This affected the entire organization. It didn't have a clear vision which led to no direction.

1. Constantly beating on subordinate leaders is counter-productive. 2. Give guidance, clear guidance with expectations and get out of the way.

You can't necessarily treat all people the same way. People respect hard but fair leaders in the long run.

Bad or untimely decisions affect lower levels exponentially (Bad). Attitude both positive and negative plays a large part in the success or failure of commanders.

Ideas-based leadership was more effective than "my way or no way" leadership.

In time of war make sure an RFI [request for information] contributes to the overall mission and not personal curiosity.

Bad—learned from a bad leader that not communicating with subordinates creates a bad work situation. Good—listening is very important.

Subordinates need to have interaction with their leaders. As a Company Commander, my Brigade Commander had very little interaction with his commanders.

Effective listening. Involve subordinates in planning process, they more often than not have great ideas and concepts, training events. The ability to accept ideas from others as a leader will enhance you and your team's effectiveness.

Positive attitudes are contagious, so are negative. Confidence while briefing and on radio goes a long way. Do not waste soldiers' time. If training is complete they should be sent home (not waiting for a higher meeting to finish).

Know your history, lessons learned will be re-learned if not understood or known the first time. Be consistent and a man of your word, failure here will have soldiers questioning your loyalty.

Learned that no matter what, soldiers respond to positive leadership; but you need to be sincere with them. Soldiers can see right through you if you are not sincere. Bad leaders are those that are selfish and perform for their careers and not their soldiers.

I have learned that COs do not go through the staff to see their boss. I have learned that the staff's job is to support the companies and the CO and sometimes must be pushed to do so. I have learned that a commander must move in his AO [area of operation] with his troops to be effective at the company level.

Good: know your soldiers, put yourself in their place when your soldiers are working the hardest, spend time with soldiers, be accessible, enforce training to standard, and be tough but fair when it comes to punishments. Bad: never abuse your power, your career is secondary to the welfare of your soldiers, and take the butt chewing if it means better results for your soldiers.

Discipline and training are paramount. The best way to take care of a soldier is to ensure he is prepared for the challenges of combat. We are currently fighting what the Marines call the 3 block war and soldiers need to be able to transition from humanitarian support to peacekeeping to High Intensity Ops on a second's notice.

1) In reference to negative counseling, the chain of command is disinterested in your reasoning of the circumstances. 2) They don't try to relate with the soldiers or subordinates; just basically toss out a bunch of orders at you. They don't or some don't take the time to develop junior leaders.

Keeping people informed and looking out for their best interest is important. You must display tactical and technical proficiency combined with common sense to make subordinates want to follow you. You must take criticisms and use others' experiences to make decisions. You must be decisive, nobody likes to follow people who waiver about decisions.

From the bad examples I learned to keep leaders well informed, always have a plan, listen to subordinates, and not to belittle people in front of their peers.

I learned to treat soldiers as I would like to be treated. Do not use too many quotes—be original; it is good to be smart but better to be a smart and effective communicator. Never belittle subordinates in front of soldiers.

Be fair, maintain control, instill trust, mentor junior leaders, train hard, allow mistakes to happen but make sure it gets corrected, leaders need to learn "situational leadership"—each person is different and will react differently depending on how they are treated. For example if you yell at one person they fix their error, yell at someone else and they will shut down.

Communication is key—2-way flow of information increases mission accomplishment, improves morale because lower enlisted know what's going on. Without decisiveness, the unit flounders and ends up wasting time. Company commanders have to be on top of all decisions.

Very poor leadership—could never make a decision either way, and in the rare cases when a decision was made, it was made out of fear of battalion commander. Also, he would never get angry, even if it was completely justified, leading to a failure to hold people accountable.

I have learned numerous lessons—too much to list. However, the most apparent answer is that sincerity means everything. If a person is self-seeking they are bad, if not they are good. What is worse than all of this is the person who is a "yes" man, who doesn't have the nerve to stand against something when they know that it is wrong.

Integrity is non-negotiable, and competence is the number one attribute of a leader. Leaders must be willing to listen and learn from others, and must be effective communicators.

Good garrison commanders are not necessarily good combat commanders. Two things matter in a combat leader: tactical genius and inflexible morals. A thousand other qualities make you look good "in the rear."

Effective communication is more than just talking and listening. It is essential to choose the right words and to be absolutely honest (knowledge, capabilities, etc.) in the presentation. Don't hide from mistakes, acknowledge and fix them, move on.

Good—Mentorship, taking time to professionally develop subordinate leaders regardless of rank. Bad—Integrity, once you compromise your integrity you can never gain it back. Subordinates will figure out you lied/cheated and they will talk.

Leadership to me is whether you can get your men to climb to the top of the mountain no matter what the conditions. Bad leaders cross the line when they don't know what they should know while they are in a leadership position, i.e., tactics, physical fitness, job knowledge.

Based on personal experience I have learned that as leaders we cannot rush to judge soldiers before we know their capabilities and limits. In turn, being in touch with your men will allow you to effectively plan and execute your mission. I've had leaders that were involved and in tune with the troops and I've had leaders that were the opposite. To be successful you have to find the balance but also do your best to get rid of stigma because the soldiers in today's Army will rise to the challenge they are presented with.

Leaders that do not defend or speak up when their subordinates are being ridiculed. Leaders that do not recognize their subordinates for the work accomplished. Leaders that are readily available to answer/listen to any and all concerns. Peers that have more knowledge and one professional enough to provide assistance.

Good: Training and educating soldier on warfighting. OPDs. Emphasis on tactical decisions and learning doctrine. Bad: Placing too much emphasis on administrative tasks. Flying off the handle at small mistakes. Focusing on NCO activities instead of officer training.

Good: Good leaders teach/coach/mentor. Delegate and praise when get good results. Be yourself. Bad: Condescending, arrogant, self-important, self-promoting superiors who lack interpersonal skills.

Taking ownership for actions both good and bad. Inflexible on rules when they don't make sense. Not taking care of soldiers.

Some officers care only about looking good for their higher commander as opposed to actually being at their jobs, they did not care about missions or subordinates.

Good—Be involved with your subordinates. Develop the team around you. Make sure all are synched on what your goals are as an organization. Bad—People focus more on professional development (i.e., OERs, schools, etc.) and less on the pure and simple basics found in the warrior ethos. Shoot, move, communicate.

Battalion commander (poor example)—did not stand up for his people (I was a company commander) especially senior leaders. I wrote him off as a mentor. I reduced my contact with him—with all due respect—to absolute minimum. Company 1SG [first sergeant], other NCOs (good examples)—I strove to fulfill their assignments and make them look good. I cleared the way—as company commander—so the 1SG had all the leeway possible to run his leadership. This means I also went to bat for him by keeping the battalion CSM and battalion commander away as much as possible.

Bad leadership has always emulated a self-serving agenda and subordinates often see and recognize it quickly. Good leadership always looks at "did I do the most as a leader to set the conditions for my subordinates' success." They look inward first, then at others.

It is important for leaders to understand their soldiers for what they are, not what the leader thinks they are/should be. Self-serving leaders are easy to identify. Good leaders listening. They don't attempt to justify. They really listen. Hard work does not necessarily mean good leadership. Leaders need to understand how their actions are perceived, not how they intended them.

Bad—First phone call to me as a company commander in Afghanistan 2001 was my XO wondering why my IMPAC credit card had not been certified and he was going to get a red bubble on some QTB [quarterly training brief] chart. He had no perspective that I was in combat and I lost faith that he was focused on supporting me forward. Good—MY 1SG was the calmest guy. He made any situation more solvable just by his presence. He had an authority that was unspoken. Helped me be a better commander and keep my sanity and perspective!

I learned how important it is to be able to access the functionality of a staff and to reorganize it as necessary (through hiring and firing process) in order to maximize its capability. Just one person not pulling his/her weight on a 4-person staff reduces its efficiency by 25% and increases the work of the other 3.

Bad—Don't talk bad about your subordinates in public, even if in jest. You have to praise hard work more than you correct shortcomings. If a commander consistently focuses on how the team can improve or do better without providing praise/acknowledgement for work done the unit will become demoralized and no longer strive to succeed.

"Good" leaders possess the qualities listed above [in response to question #1 asking for characteristics of a leader the officer admires, the officer listed competent, thorough, patient, loyal, non-competitive, motivated, optimistic about future, supports chain of command]. "Bad" leaders are more concerned with their self-preservation, getting credit for their accomplishment (must tell someone), focus only on their strengths not their weaknesses. Evaluate the success of others relative to their own.

APPENDIX D

# Sample Battalion Commander Development Form

One of the battalion commanders we met with created this document and used it as a developmental form for junior officers in her unit. It is one example of many that senior leaders told us they had created on their own for use by their unit. It references official Army counseling forms and requires their use while providing guidance on how to counsel junior officers in the unit. It is presented here only as an example of the types of resources and tools that are already present in the field and that might be shared with others. We are not recommending that this particular form be required or that all battalion commanders should maintain such detailed information on each of their junior officers.

## BACKGROUND DATA

**Promotion Data**

1. DOR -
2. BZ Board -
3. PZ Board -

**PMOS**

1. Branch - Qualified in Grade - YES / NO (Refer to DA PAM 600-3)
2. Secondary -

**Schools**

1. CSC - YES / NO School Attended - USA / USAF / USMC / USN / Other Date -
2. JPME II - YES / NO Date -
3. Masters Degree Awarded - Date -
4. PHd Awarded - Date -
5. NSDP Date - 2000

**Previous Positions Held** (Does the Officer have a balanced Tactical/Strategic Path)

1. XO
2. S3
3. Company Commander
4. S2

**Deployment Time** (If Officer does not have a current deployment this data will allow you to work with Branch or through higher S3 for WIAS Tasker upon completion of BQ job)

1. OEF 05/08
2. OIF II
3. Bosnia 1998

# RATING PERIOD COUNSELING NOTES

**DA FORM 67-9-1 Support Form** (Provide within 30 days of assignment to current duty position)

1. Provide officer copy of Senior Rater support form
2. Provide officer copy of Rater support form

**Individual DA 67-9-1**

1. A living contract, changes should be annotated at quarterly counseling on the form
2. The -1 is a guide to what you both believe the officer should be able to achieve
3. Should be completed within 30 days of assuming duty position
4. Should address major objectives for the individual duty position - company commander will be significantly different from operations officer
5. All items performance objectives should fit on the form; take guidance from your Senior Rater on the length of this document; I've had up to 20 additional pages that one SR wanted and my last SR wanted your top 3 accomplishments
6. Per CSA Guidance it must address SAFETY objectives/performance
7. I believe in the one year rating period that you will formally counsel the individual at least 6 times -
   a. Initial + 30 Days
   b. $1^{st}$ QTR
   c. $2^{nd}$ QTR
   d. $3^{rd}$ QTR
   e. $4^{th}$ QTR - 10-11 months - request completed form NLT beginning of $11^{th}$ month
   f. OER counseling at end of rated period
8. If you have to counsel the officer on any other occasion use a Memorandum format. This requires 2 memorandums. The first is the counseling with the FOR line addressed to the officer which needs their endorsement at the bottom after yours - you keep the original; the second is a memorandum for record for your files.

**RATING PERIOD 1 (first 30 days)**

**Rated Officer Initials:** **Rater Initials:** **Date:**

**Notes**

1. Review -1. Tell the Officer exactly what you expect.
2. Make adjustments directly on the form. Have the officer correct any changes then sign and date. You should keep the original.
3. Send a copy of the -1 to the Senior Rater for initials, date
4. Tell the Officer exactly what level of authority they have - they can sign for you; spend BN money; are #3 in the Chain of Command; Speak for you; etc

**Discussion Points** I use the Army "LDRSHIP" acronym as my guide for discussion points for the initial counseling only.

1. Loyalty - to self; Soldiers; unit; mission.

2. Duty - Discuss what it means to be a Professional Soldier.

3. Respect - goes both ways, up/down. Ensure the officer understands Enlisted/NCOs and how they should be treated.

4. Selfless Service - Not martyrdom but ensuring the mission gets accomplished while balancing family, personal time.

5. Honor - in everything. Be honest with self, seniors, subordinates.

6. Integrity - you get one set, one chance, normally cannot be redeemed during this duty assignment if you violate it.

7. Personal Courage - to speak up and tell good, bad, ugly. Bad news does not get better, conversely good news of Soldier accomplishment should be shared immediately even if we don't get credit for the success

**RATING PERIOD 2 ($1^{st}$ QTR)**

**Rated Officer Initials:** **Rater Initials:** **Date:**

**Discussion Points**

1. Review -1. Make adjustments directly on the form. Sign and date the original and provide an updated copy to the Individual.
2. Discuss duty performance
3. Discuss timeliness issues (personal, suspenses)
4. Review education goals
5. Review personal fitness goals
6. Review balance (Army - Family - Self)
7. Send a copy of the -1 to the Senior Rater for initials, date

[NOTE: The battalion commander's document includes separate pages for the second-, third-, and fourth-quarter counseling sessions; those pages are not presented here because their format is identical to the one seen above.]

## OER Counseling - Rated Period Complete

**Rated Officer Initials: Rater Initials: Date:**

**Discussion Points**

1. Review Completed OER. Have the officer Sign. (Submission timelines +90 days to HRC for AD/ 180 days for RC)
2. Tell the officer what they did well, what they need to continue to work on.
3. Tell the officer what they should strive to achieve in their next job or rating cycle.
4. Send a copy of the OER and suggested Senior Rater comments to the SR for completion.

**OER Notes**

APPENDIX E

# Brief Review of Other Studies of Leader Development

The leadership literature is voluminous. At the end of 2006, the Library of Congress had nearly 1,400 documents with the word *leadership* in the title; a keyword search of the term found more than 10,000 documents. There are scholarly journals devoted entirely to the topic of leadership; other journals have special issues focused on leadership; leadership institutes have been established at universities, nonprofit organizations, and global corporations; and some consulting firms have a leadership practice while others focus solely on leadership. Although studies of leaders and leadership can be found in ancient texts, interest in the subject has exploded in recent years. Of the Library of Congress documents with *leadership* in the title, nearly a quarter have been published in the last five years.

This review is not meant to be an exegesis of the vast field of leader development research, but we offer it to establish a framework for describing and understanding leader development programs in the Army's operational units. In particular, it focuses attention on the role of the unit commanders and on counseling, coaching, and mentoring of the junior officers.

## What Can Leader Development Programs Develop?

After three decades of studying leader development programs, the Center for Creative Leadership has identified a range of leader capabilities that can be developed. Broadly, they fall into three categories: self-management capabilities, social capabilities, and work facilitation capabilities.[1] Many of these capabilities are similar to elements of the Army's leadership requirements model, such as communication skills, creativity, values, developing others, and self-awareness. In other words, many of the attributes and competencies required of Army leaders can be developed. Not surprisingly, they are largely developed through experience as opposed to training or education. However, recent research by RAND Arroyo did suggest a greater role for educational organizations and distance learning in developing certain cognitive requirements for leaders—pattern recognition, perceptual acuity, mental simulation, and critical thinking.[2]

## Key Characteristics of Effective Leader Development Programs

With so many titles and publications examining leaders and leadership, some authors have attempted to identify common themes and best practices across organizations. The Army Research Institute (ARI) conducted one such study, which was published in 2001. In a different study, the Army Management Staff College was featured in *Linkage Inc.'s Best Practices in Leadership Development Handbook*. From these and other studies, a handful of key themes emerge. One is the importance of having senior leaders who strongly support leader development within their organization. The authors of the ARI study write, "It appears that the most important principle in successful leadership development efforts is the presence of an influential champion."[3] Linkage, Inc., identified "support and involvement of senior management"

---

1 Van Velsor and McCauley, 2004.

2 Henry A. Leonard et al., *Something Old, Something New: Army Leader Development in a Dynamic Environment*, Santa Monica, Calif.: RAND Corporation, MG-281-A, 2006.

3 David V. Day and Stanley M. Halpin, *Leadership Development: A Review of Industry Best Practices*, United States Army Research Institute for the Behavioral and Social Sciences, Technical Report 1111, April 2001, p. viii.

as the single most important factor underlying a successful leader development initiative.[4] Studies of best practices for a specific type of leader development—the 360 evaluation—conclude that "boss support is critical for the process itself, as well as for buy-in for the recipient's specific developmental goals stemming from the feedback."[5] On a more informal level, write researchers at the Center for Creative Leadership, "the support of one's current boss is particularly important when trying to change behaviors or learn new skills."[6]

A second theme from the literature is the need for feedback and assessment. The more than 8,000 business leaders who responded to a 2001 survey conducted by the Corporate Leadership Council rated feedback and relationship programs as the most effective type of leader development activity.[7] Based on three decades of studying leader development programs, the Center for Creative Leadership lists assessment as one of three factors that "make developmental experiences more powerful" as a means of developing leaders.[8] The feedback and assessment can be formal or informal, and may involve peers, superiors, role models, and coaches.[9]

The emphasis on feedback extends more broadly than the leader development literature. Work experience has been found to have more developmental value when individuals have performance management and feedback systems.[10] Studies of experts and expertise consistently find that teachers and coaches play a critical role in providing feedback

---

4 Louis Carter, David Giber, and Marshall Goldsmith, eds., *Linkage Inc.'s Best Practices in Leadership Development Handbook,* San Francisco, Calif.: Jossey-Bass/Pfeiffer, 2000, p. 445.

5 Gina Hernez-Broome and Richard L. Hughes, "Leadership Development: Past, Present, and Future," *Human Resource Planning,* vol. 27, No. 1, 2004.

6 Van Velsor and McCauley, 2004, p. 11.

7 Corporate Leadership Council, *Voice of the Leader: A Qualitative Analysis of Leadership Bench Strength and Development Strategies,* Washington, D.C.: Corporate Executive Board, 2001.

8 Van Velsor and McCauley, 2004, p. 3. The other two factors are challenge and support.

9 Phillips and Schmidt, 2004, pp. 9–11.

10 Paul E. Tesluk and Rick R. Jacobs, "Toward an Integrated Model of Work Experience," *Personnel Psychology,* Vol. 51, Is. 2, June 1998, pp. 321–355.

that enables an individual to monitor and evaluate his or her performance.[11] One of the specific ways in which feedback benefits the recipient is by reducing the discrepancy between self-assessment and how others assess the individual. Closer alignment of self- and subordinate assessment is correlated with higher performance and earlier promotion of military officers.[12]

One reason why some leader development activities were ranked low in effectiveness by the junior officers may be that they are treated as isolated events rather than an ongoing process of development. From the Center for Creative Leadership, a "major criticism of the approach of many organizations to leader development is that it is not systemic but events-based."[13] The many types of activities discussed in this document, from staff rides to reading lists to officer evaluations, must fit together in a coherent way to be seen by the individual as promoting development. Even something as critical as assessment can have diminished potency if it is not treated as an ongoing process of leader development. Again, from the Center for Creative Leadership, "Multi-rater feedback is an effective assessment activity, an experience that helps unfreeze people and prepares them to learn from other developmental experiences. But if you just give someone feedback from an instrument and stop there, little real development takes place."[14] When the OER support form is completed shortly before the OER, not only is it of limited use, but the junior officer is also implicitly being taught that tools that could be used for development are really part of an event—in this case, fulfilling an administrative requirement on time.

---

11 See, for example, Ericsson, 2006.

12 Bernard. M. Bass and F.J. Yammarino, "Congruence of Self and Others' Leadership Ratings of Naval Officers for Understanding Successful Performance," *Applied Psychology: An International Review*, Vol. 40, 1991, pp. 437–454.

13 Van Velsor and McCauley, 2004, p. 18.

14 Van Velsor and McCauley, 2004, p. 18.

## Prior Army Reviews of Leader Development

Many leader development studies have been conducted by and for the Army. In the past two decades, among the more prominent are the Professional Development of Officers Study (PDOS), completed in 1985; a leader development study led by then-MG Gordon Sullivan, completed in 1988; and the Army Training and Leader Development Panel (ATLDP), completed in phases from 2001 to 2003. Each was chartered by the Chief of Staff of the Army.

The PDOS and Sullivan studies largely focused on education and training, such as the timing of school attendance in a career, course curricula, and military qualification standards. But both also addressed the issues of self-development and mentoring. Among the PDOS findings were that "too many officers perceive they do not have mentoring leaders and schools do not contribute as effectively as they might in this regard."[15] The PDOS called for a "mentorship-based method of instruction" at educational institutions and in units.[16] This required that "commanders acknowledge and claim responsibility for the mentorship role, adopt experiential learning/teaching methodologies, and establish the necessary development climate within which constructive feedback is provided to the individual officer."[17] The Sullivan study noted the importance of feedback during counseling sessions and AARs as critical to the development of leaders.[18] Similar to the new research presented in this monograph, the Sullivan study noted that there were "indications . . . that there is considerable disparity in the quality of OPD programs throughout the force." The study did not elaborate.[19]

---

15 Headquarters, Department of the Army, *Professional Development of Officers Study: Final Report for the Chief of Staff, Army, Volume I*, Washington, D.C., 1985, p. 33.

16 Headquarters, Department of the Army, *Professional Development of Officers Study, Final Report for the Chief of Staff, Army, Volume III*, Washington, D.C., 1985, p. I-4.

17 Headquarters, Department of the Army, Volume III, 1985, p. I-4.

18 Headquarters, Department of the Army, 1988, p. 10.

19 Headquarters, Department of the Army, 1988, p. 10.

The ATLDP arguably had the broadest focus of the aforementioned studies. One of the key conclusions of the effort was that "the Army must have standards and effective assessment, evaluation, and feedback systems for leaders, units, and itself," but noted that "there appears to be no approved feedback mechanism for individual leaders."[20] Furthermore, the study noted that the OER is not being used as a leader development tool, and senior raters seldom counsel subordinates.[21] The study found that officers do not want formal mentoring programs, just as the current research did, but did conclude that officers want Army culture to put a greater emphasis on mentoring.[22] Recommendations included "reinforcing the leader development aspects of the OER to increase communications between junior and senior officers" and emphasizing mentoring in Pre-Command Courses to future battalion and brigade commanders.[23] On the issue of self-development, the study concluded that it does not receive adequate emphasis from Army leaders and that officers lack the tools and support for self-development.[24]

---

20 U.S. Army Combined Arms Center, 2001, p. OS-3.

21 U.S. Army Combined Arms Center, 2001, p. OS-9.

22 U.S. Army Combined Arms Center, 2001, p. OS-10.

23 U.S. Army Combined Arms Center, 2001, p. OS-11.

24 U.S. Army Combined Arms Center, 2001, p. OS-18.

# References


Bass, Bernard M., and F. J. Yammarino, "Congruence of Self and Others' Leadership Ratings of Naval Officers for Understanding Successful Performance," *Applied Psychology: An International Review*, Vol. 40, 1991.

Carter, Louis, David Giber, and Marshall Goldsmith, eds., *Linkage Inc.'s Best Practices in Leadership Development Handbook,* San Francisco, Calif.: Jossey-Bass/ Pfeiffer, 2000.

Corporate Leadership Council, *Voice of the Leader: A Qualitative Analysis of Leadership Bench Strength and Development Strategies*, Washington, D.C.: Corporate Executive Board, 2001.

Day, David V., "Leadership Development: A Review in Context," *Leadership Quarterly*, Vol. 11, No. 4, 2000.

Day, David V., and Stanley M. Halpin, *Leadership Development: A Review of Industry Best Practices*, United States Army Research Institute for the Behavioral and Social Sciences, Technical Report 1111, April 2001.

Ericsson, K. Anders, "The Influence of Experience and Deliberate Practice on the Development of Superior Expert Performance," in K. Anders Ericsson, Neil Charness, Robert R. Hoffman, and Paul J. Feltovich, eds., *The Cambridge Handbook of Expertise and Expert Performance*, Cambridge, UK: Cambridge University Press, 2006.

Gardner, John, *On Leadership*, New York: Free Press, 1989.

Headquarters, Department of the Army, *Professional Development of Officers Study: Final Report for the Chief of Staff, Army, Volume I,* Washington, D.C., 1985.

———, *Professional Development of Officers Study: Final Report for the Chief of Staff, Army, Volume III,* Washington, D.C., 1985.

———, *Leader Development Study: Final Report*, Washington, D.C., 1988.

———, *Leader Development for America's Army*, Pamphlet 350-58, Washington, D.C., October 1994.

———, *Battle Focused Training*, Field Manual 7-1, Washington, D.C., September 2003.

———, *Evaluation Reporting System*, AR 623-3, Washington, D.C., May 2006.

———, *Army Leadership: Competent, Confident, and Agile*, Field Manual 6-22, Washington, D.C., October 2006.

Hernez-Broome, Gina, and Richard L. Hughes, "Leadership Development: Past, Present, and Future," *Human Resource Planning*, Vol. 27, No. 1, 2004.

Leonard, Henry A., J. Michael Polich, Jeffery D. Peterson, Ronald E. Sortor, and S. Craig Moore, *Something Old, Something New: Army Leader Development in a Dynamic Environment*, Santa Monica, Calif.: RAND Corporation, MG-281-A, 2006. As of June 27, 2007:
http://www.rand.org/pubs/monographs/MG281/

Lewis, Mark R., "Army Transformation and the Junior Officer Exodus," *Armed Forces and Society*, Vol. 31, No. 1, Fall 2004, pp. 63–93.

McCausland, Jeffrey D., "Observations, Conclusions, and Recommendations," in Jeffrey D. McCausland, ed., *Educating Leaders in an Age of Uncertainty: The Future of Military War Colleges*, Carlisle, Pa.: Dickinson College, 2005.

Myrer, Anton, *Once an Eagle*, New York: Harper Collins, 2001.

Phillips, Jack J., and Lynn Schmidt, *The Leadership Scorecard: ROI for Leaders (Improving Human Performance Series)*, Burlington, Mass.: Elsevier Butterworth-Heinemann, 2004.

Scandura, Terri A., and Chester A. Schriesheim, "Leader-Member Exchange and Supervisor Career Mentoring as Complementary Constructs in Leadership Research," *Academy of Management Journal*, Vol. 37, No. 6, 1994.

Senge, Peter M., *The Fifth Discipline*, New York: Doubleday/Currency, 1990.

Snider, Don M., "The U.S. Army as Profession," in Don M. Snider and Lloyd J. Matthews, eds., *The Future of the Army Profession*, 2nd ed., Boston, Mass.: McGraw-Hill Custom Publishing, 2005.

Snider, Don M., and Lloyd Matthews, eds., *The Future of the Army Profession*, 2nd ed., Boston, Mass.: McGraw-Hill Custom Publishing, 2005.

Tesluk, Paul E., and Rick R. Jacobs, "Toward an Integrated Model of Work Experience," *Personnel Psychology*, Vol. 51, Is. 2, Summer 1998.

Ulmer, Walter F., Jr., Michael D. Shaler, R. Craig Bullis, Diane F. DiClemente, and T. Owen Jacobs, *Leadership Lessons at Division Command Level—2004*, Carlisle Barracks, Pa.: U.S. Army War College, 2004.

U.S. Army Combined Arms Center, *Army Training and Leader Development Panel Officer Study Report to the Army*, Fort Leavenworth, Kan., 2001.

Van Velsor, Ellen, and Cynthia D. McCauley, eds., *The Center for Creative Leadership Handbook of Leadership Development*, 2nd ed., San Francisco, Calif.: Jossey-Bass Publishers, 2004 .

Wong, Leonard, *Generations Apart: Xers and Boomers in the Officer Corps*, Carlisle Barracks, Pa.: U.S. Army War College, Strategic Studies Institute, 2000.